MEDICAL, DENTAL, AND WELLNESS TOURISM

A Post-Pandemic Perspective

Advances in Hospitality and Tourism Series

MEDICAL, DENTAL, AND WELLNESS TOURISM

A Post-Pandemic Perspective

Mary Schreiber Swenson, PhD

Amit Bansal, COAUTHOR

with contributions from Sunidhi Raiya

AAP | APPLE ACADEMIC PRESS

First edition published 2024

Apple Academic Press Inc.
1265 Goldenrod Circle, NE,
Palm Bay, FL 32905 USA

760 Laurentian Drive, Unit 19,
Burlington, ON L7N 0A4, CANADA

CRC Press
2385 NW Executive Center Drive,
Suite 320, Boca Raton FL 33431

4 Park Square, Milton Park,
Abingdon, Oxon, OX14 4RN UK

Library and Archives Canada Cataloguing in Publication

Title: Medical, dental, and wellness tourism : a post-pandemic perspective / Mary Schreiber Swenson, PhD, Amit Bansal, coauthor ; with contributions from Sunidhi Raiya.
Names: Swenson, Mary Schreiber, author. | Bansal, Amit, author.
Series: Advances in hospitality and tourism book series.
Description: First edition. | Series statement: Advances in hospitality and tourism book series | Includes bibliographical references and index.
Identifiers: Canadiana (print) 20230445578 | Canadiana (ebook) 20230445608 | ISBN 9781774914946 (hardcover) | ISBN 9781774914953 (softcover) | ISBN 9781003433569 (ebook)
Subjects: LCSH: Medical tourism. | LCSH: Medical tourism—Technological innovations.
Classification: LCC RA793.5 .S94 2023 | DDC 362.1—dc23

Library of Congress Cataloging-in-Publication Data

CIP data on file with US Library of Congress

ISBN: 978-1-77491-494-6 (hbk)
ISBN: 978-1-77491-495-3 (pbk)
ISBN: 978-1-00343-356-9 (ebk)

ABOUT THE ADVANCES IN HOSPITALITY AND TOURISM BOOK SERIES

Editor-in-Chief:
Mahmood A. Khan, PhD
Professor, Department of Hospitality and Tourism Management,
Pamplin College of Business,
Virginia Polytechnic Institute and State University,
Falls Church, Virginia, USA
Email: mahmood@vt.edu

This series reports on research developments and advances in the rapidly growing area of hospitality and tourism. Each volume in this series presents state-of-the-art information on a specialized topic of current interest. These one-of-a-kind publications are valuable resources for academia as well as for professionals in the industrial sector.

BOOKS IN THE SERIES:

Food Safety: Researching the Hazard in Hazardous Foods
Editors: Barbara Almanza, PhD, RD, and Richard Ghiselli, PhD

Strategic Winery Tourism and Management: Building Competitive Winery Tourism and Winery Management Strategy
Editor: Kyuho Lee, PhD

Sustainability, Social Responsibility and Innovations in the Hospitality Industry
Editor: H. G. Parsa, PhD
Consulting Editor: Vivaja "Vi" Narapareddy, PhD
Associate Editors: SooCheong (Shawn) Jang, PhD,
Marival Segarra-Oña, PhD, and Rachel J. C. Chen, PhD, CHE

Managing Sustainability in the Hospitality and Tourism Industry: Paradigms and Directions for the Future
Editor: Vinnie Jauhari, PhD

Management Science in Hospitality and Tourism: Theory, Practice, and Applications
Editors: Muzaffer Uysal, PhD, Zvi Schwartz, PhD, and
Ercan Sirakaya-Turk, PhD

Tourism in Central Asia: Issues and Challenges
Editors: Kemal Kantarci, PhD, Muzaffer Uysal, PhD, and Vincent Magnini, PhD

Poverty Alleviation through Tourism Development: A Comprehensive and Integrated Approach
Robertico Croes, PhD, and Manuel Rivera, PhD

Chinese Outbound Tourism 2.0
Editor: Xiang (Robert) Li, PhD

Hospitality Marketing and Consumer Behavior: Creating Memorable Experiences
Editor: Vinnie Jauhari, PhD

Women and Travel: Historical and Contemporary Perspectives
Editors: Catheryn Khoo-Lattimore, PhD, and Erica Wilson, PhD

Wilderness of Wildlife Tourism
Editor: Johra Kayeser Fatima, PhD

Medical Tourism and Wellness: Hospitality Bridging Healthcare (H_2H)©
Editor: Frederick J. DeMicco, PhD, RD

Sustainable Viticulture: The Vines and Wines of Burgundy
Claude Chapuis

The Indian Hospitality Industry: Dynamics and Future Trends
Editors: Sandeep Munjal and Sudhanshu Bhushan

Tourism Development and Destination Branding through Content Marketing Strategies and Social Media
Editor: Anukrati Sharma, PhD

Evolving Paradigms in Tourism and Hospitality in Developing Countries: A Case Study of India
Editors: Bindi Varghese, PhD

The Hospitality and Tourism Industry in China: New Growth, Trends, and Developments
Editors: Jinlin Zhao, PhD

Labor in Tourism and Hospitality Industry: Skills, Ethics, Issues, and Rights
Abdallah M. Elshaer, PhD, and Asmaa M. Marzouk, PhD

Sustainable Tourism Development: Futuristic Approaches
Editor: Anukrati Sharma, PhD

Tourism in Turkey: A Comprehensive Overview and Analysis for Sustainable Alternative Tourism
Editor: Ahmet Salih İKİZ

Capacity Building Through Heritage Tourism: An International Perspective
Editor: Surabhi Srivastava, PhD

Medical Travel Brand Management: Success Strategies for Hospitality Bridging Healthcare (H2H)
Editors: Frederick J. DeMicco, PhD, RD, and Ali A. Poorani, PhD

Strategic Management for the Hospitality and Tourism Industry: Developing a Competitive Advantage
Editor: Vincent Sabourin, PhD

Tourist Behavior: Past, Present, and Future
Editors: Narendra Kumar, PhD, Bruno Barbosa Sousa, PhD, and Swati Sharma, PhD

The Hospitality and Tourism Industry in ASEAN and East Asian Destinations: New Growth, Trends, and Developments
Editors: Jinlin Zhao, PhD, Lianping Ren, D.HTM, and Xiangping Li, PhD

Event Tourism in Asian Countries: Challenges and Prospects
Editor: Shruti Arora, PhD

Small Island and Small Destination Tourism: Overcoming the Smallness Barrier for Economic Growth and Tourism Competitiveness
Robertico Croes, PhD

Post-COVID Tourism and Hospitality Dynamics: Recovery, Revival, and Re-start
Editors: Umendra Narayan Shukla, PhD, and Sharad Kumar Kulshreshtha, PhD

Brand Co-Creation Tourism Research: Contemporary Issues and Challenges
Editor: Raouf Ahmad Rather, PhD

Dynamics of the Tourism Industry: Post-Pandemic and Post-Disaster Perspectives and Strategies
Editors: Gül Erkol Bayram, PhD, and Anukrati Sharma, PhD

Medical, Dental, and Wellness Tourism: A Post-Pandemic Perspective
Mary Schreiber Swenson, PhD, and Amit Bansal

ABOUT THE SERIES EDITOR

Mahmood A. Khan, PhD, is a Professor in the Department of Hospitality and Tourism Management, Pamplin College of Business at Virginia Tech's National Capital Region campus. He has served in teaching, research, and administrative positions for the past 35 years, working at major US universities. Dr. Khan is the author of several books and has traveled extensively for teaching and consulting on management issues and franchising. He has been invited by national and international corporations to serve as a speaker, keynote speaker, and seminar presenter on different topics related to franchising and services management. Dr. Khan has received the Steven Fletcher Award for his outstanding contribution to hospitality education and research. He is also a recipient of the John Wiley & Sons Award for lifetime contribution to outstanding research and scholarship; the Donald K. Tressler Award for scholarship; and the Cesar Ritz Award for scholarly contribution. He also received the Outstanding Doctoral Faculty Award from Pamplin College of Business. He has served on the Board of Governors of the Educational Foundation of the International Franchise Association, on the Board of Directors of the Virginia Hospitality and Tourism Association, as a Trustee of the International College of Hospitality Management, and as a Trustee on the Foundation of the Hospitality Sales and Marketing Association's International Association. He is also a member of several professional associations.

ABOUT THE AUTHORS

Mary Schreiber Swenson, PhD, is the CEO and Founder of MyMedChoices™, the global leader in providing surgical, dental, and wellness international solutions. She has traveled the globe in search of value-based Centers of Excellence and healthcare solutions for individuals, employers, and government agencies to ensure the highest quality of care and transparent pricing.

Acknowledged for her unique position as a female CEO, Dr. Swenson earned membership to the Harvard University Kennedy School's Women's Leadership Board following her success bringing two public companies—GUAVA Healthcare and GUAVA Franchising—into Entrepreneur Magazine's Top Franchise 500 for four years running. There, Dr. Swenson plays a crucial role with the Women's Leadership Board in amplifying research and transforming the world toward gender equity through key partnerships with world leaders, a topic explored in her previously published work, *Women Economic Empowerment and Success*.

Most recently, Dr. Swenson was awarded the 2021 Enterprising Women of the Year Award, the most prestigious recognition for business owners around the world, and was later inducted into the Hall of Fame on November 16, 2021. Other accolades include the President's Lifetime Achievement Award, given to Dr. Swenson by the Office of the President of the United States and AmeriCorps in December 2021.

Always on the cutting edge of wellness trends, Dr. Swenson founded Vegan Skin Clinic, an organic, plant-based skincare line creating a new awareness for natural wellness solutions and a "brainier beauty," during the COVID-19 pandemic.

Amit Bansal is a technology veteran with over 20 years of experience in conceptualizing and implementing large enterprise systems in a variety of industries, including healthcare. Amit has traveled to 24+ countries and worked with some of the large companies and helped them significantly improve their process efficiency. Being on both sides of the business, Amit has observed how healthcare is perceived and delivered worldwide and his own experience in 2014 made him incept MediGence with the vision to improve global healthcare access. His goal is to take the best healthcare in every part of the world and enable people to access it anytime, anywhere by paying a minimum monthly subscription fee.

He utilizes technology to make communication between patients and physicians smooth and help them make better diagnoses and treatment decisions. His strengths are the ability to multitask, passion for work, and strong work ethic. Amit strongly believes in the goodness of people and has a natural tendency to help souls. He enjoys interacting with people, learning new things, traveling, and exploring new creative ideas.

With contributions from:

Sunidhi Raiya is a healthcare enthusiast who is constantly evolving and investing herself as a voice in the healthcare community. She has years of experience in healthcare research. Her goal is to bring change and make the world a better place through her small efforts.

CONTENTS

ABBREVIATIONS

ACOs	accountable care organizations
CoE	Centers of Excellence
EHR	electronic health record
ICTs	information and communication technologies
IR	Industrial Revolution
PCMHs	patient-centered medical homes
PPE	personal protective equipment
SLT	speech and language therapy

FOREWORD

As a former CEO of one of Australia's largest private hospitals for 10 years, plus more than 40 years managing hospitals and healthcare facilities in both the public and private sectors in four Australian states, I consider myself to be an experienced hospital and healthcare manager. Prior to COVID-19, my employing company, a global private hospital and healthcare provider of considerable size and reputation, asked me to develop a medical tourism business for our Australian hospitals—essentially to provide an additional revenue source at a time when the traditional revenue from private health funds was in decline. A number of Australian private hospitals, including my own, had previously "dabbled" in medical tourism. It wasn't until Mary Schreiber Swenson, founder and owner of MyMedChoices, met with my company executives and effectively painted a picture of what a medical tourism business could look like for Australia and the enormous potential that it represented, and what dedicated resources, such as myself, allocated to the concept.

Medical Tourism, Medi-tourism, and Travel for Treatment are all descriptors for travel primarily for the purpose of receiving healthcare—not a new concept, but one that's been gaining momentum, in particular in countries that have a standard and quality of healthcare that is considered desirable to those seeking treatment, like cost-effectiveness, access to better treatments, technologies and credentialed doctors, techniques, as well as better facilities, heightened quality and safety standards, and the opportunity to combine medical care with leisure tourism opportunities.

As an Australian (and like most Australians), I understandably tend to look at the world as it relates to Australia. Medical tourism has had an identity crisis, at least in Australia, in that it has negative stereotypes and stigmas associated with it, caused by the narrow misconception that it's comprised solely of cosmetic procedures, e.g., breast enlargements and teeth whitening, performed at cheaper rates and often in third-world countries—a far cry from the modernized standards of the contemporary medical tourism industry.

The reality of medical tourism and travel for treatment is that patients are prepared to travel all over the world to seek all manners of surgical interventions, such as joint replacements, cancer surgery, and neurosurgery from top doctors and state-of-the-art institutions at a fraction of the cost offered by healthcare systems such as the United States. I remember being surprised when discussing medical tourism and travel for treatment, particularly as it related to the United States and Australia, with a friend who is an ICU physician from Maryland, especially when hearing that the cost of a night in ICU in his US hospital was three times the cost of a night in ICU of my hospital. Scenarios like this put the whole concept of medical tourism and travel for treatment, and why people want to feel empowered to make their own treatment choices, into perspective.

Upon its arrival, the COVID-19 pandemic temporarily, but effectively, stalled the forward progress the medical tourism industry had made in many countries, particularly in my native Australia, which effectively used its absence of shared borders to isolate itself from the rest of the world. For much of the pandemic, Australia was closed to the rest of the world with very few people being allowed in or out—not an environment that is necessarily conducive to an expanding international medical tourism industry. But outside of Australia, many other countries chose a different approach to managing the pandemic, including a far more liberal approach to border access and, subsequently, in many of these countries medical tourism continued pandemic-wide and actually thrived, while more domestic travel for treatment opportunities took center stage in isolated countries like Australia. Nevertheless, the overall impact of the pandemic was that it deprived the world of the opportunity to develop and expand the concept of medical tourism at a time when global interest was growing rapidly.

Dr. Swenson had opened our eyes to a potential industry that had effectively been lying dormant in Australia although until the pandemic had flourished in many countries around the world. In my newly appointed role, I began discussions with Dr. Swenson as well as several other organizations representing potential patients and referring organizations in China, New Zealand, UAE, India and, of course, the USA. It was during the course of these discussions that I started to appreciate the full magnitude of medical tourism and travel for treatment globally, and with Mary's guidance and encouragement, I began to understand the true potential it had for Australia and the rest of the world.

I had previously been aware of individuals within small organizations in Australia who, collectively and individually, had attempted to convince successive Australian governments that medical tourism in its various forms and descriptors was a legitimate industry with huge upside for Australia. They had gained little traction. It presented an attractive challenge that will be made much more achievable because of Dr. Swenson's book, which could well be the paragon work on the subject of medical tourism and travel for treatment.

Medical, Dental and Wellness Tourism: A Post-Pandemic Perspective is truly the "ready reckoner" on the subject of medical tourism and addresses the issues encountered by various countries, such as the differing approaches to patient data transfers whilst maintaining and complying with various privacy and confidentiality laws. But perhaps, most importantly, it plots the way forward, emphasizing the importance of a collaborative global approach, the rise of wellness tourism and the growing involvement and utilization of Telehealth.

—Greg Jenke, MBA
CEO, John Flynn Private Hospital,
Ramsay Health Care, Australia

PREFACE

As of now, I am claiming my power by taking full responsibility for a healthcare system that is not giving me choices and options, not giving me transparent pricing, not allowing me to travel outside my geographical area and leaving me and my family sick. All because I have not spoken up. Why, when it comes to our healthcare costs and transparency in pricing, "we the people" continue not to get angry and why don't we insist on change? Why is it that I can go into any car dealership anywhere in the United States and, after doing comparative shopping, I can pick out the car best suited to my needs, negotiate the price, and then make a purchase—all the while knowing the exact bottom-line pricing on my shiny new car? I can even walk away with a guarantee! And I don't have to worry about a surprise bill arriving two weeks later for unexpected, additional charges.

However, when I go in for a knee replacement surgery, I have no idea how much it costs. I don't have the ability to make comparisons on hospitals to find out if I am going to the best hospital with the best outcomes. I don't know anything about the facility other than what my doctor recommends or what Aunt Marge recommended over coffee last week. I don't have the ability to comparatively shop within my area or outside the United States and I have to stay in my geographical area. All topped off by the "surprise bill" I receive in the mail for a charge in the hospital that I have no idea what it is for. So, I have no idea what the overall cost or outcome is for my knee surgery. I have no idea what surprise bills I will be receiving over the next month. I have no idea if I was really at the best place with the best outcome for knee surgery or if there is a guarantee. Therefore, I could, in fact, return to the hospital beyond my control and be charged again for complications and receive yet another bill. And "we the people" are ok with this?

As of right now, I am claiming my power by taking full responsibility for the healthcare system that is keeping me sick and broke because "I" failed to speak up. When I was going through my divorce and starting a new business, I was so broke. Raising three kids and starting a business but not broke enough for Medicaid, I made just enough money to end up paying those high deductibles—which actually kept us away from the doctor—and yet, even then I never spoke up. I will never forget this next story, and I actually did not know whether to share, but here it goes…

Raising my children as a single mom and juggling the household monies, my daughter came to me with a cut on her leg, I brought her to the doctor as the cut was not healing. We were sitting in the waiting room and I looked at her and said, "I am going up to the counter to find out what I have to pay out of pocket, if it is over $200 we are going to have to make a choice on groceries or this visit. She looked at me and said, "let's leave mom." We left, and she returned a week later with MRSA. Yes, I have beaten myself up about this event every day and if I experienced this, I know you have too.

Why do I have to make these choices? Why is it easier for me to plan my Disney Trip than it is to plan my colonoscopy? I know exactly what theme park I will be visiting, the name of my pick-up person in Florida, I even plan my 6:00 pm dinner, down to the salad dressing with Mickey Mouse, a full itinerary on a play-by-play action-packed day and transparent pricing. But not with my colonoscopy and double the price. I would be lucky if my doctor even comes in after the procedure to just ask me about my overall experience.

This is not even the crazy part of this experience; what is unbelievable is that I am 54 years old and still continue this healthcare behavior. Almost as if I have accepted it and don't have a choice in the matter. I continue booking my doctor's appointments and surgery, when needed, and still have no clue as to what the cost is.

So, eight years ago, I started traveling the globe looking at treatments and options and how other countries deal with healthcare. This book is not going to be one of those "bashing the United States" books and I am not going to ask the United States to set up another healthcare system. My book is about how we are going to collectively work on changing the current US healthcare system to be a bipartisan issue. How will we, the most powerful person in the United States, start taking action?

So let's talk solutions.

—Mary Schreiber Swenson, PhD

CHAPTER 1

DEFINING MEDICAL, DENTAL, AND WELLNESS TOURISM AND TRAVEL FOR TREATMENT

ABSTRACT

While the travel for treatment industry has existed for decades, varying definitions of the term "medical tourism" itself and restrictive health care policies in consumers' home countries have turned medical and dental tourism into largely unknown concepts for many patients worldwide. This chapter explores the general facts behind modern medical tourism, including basic analyses of the US healthcare system, ways to identify accredited medical facilities, and explanations on how consumers can evaluate their options to find the best cost-to-quality ratio of care for their specific needs.

1.1 JUST THE FACTS

While medical tourism has long been touted as a solution for citizens of developing countries lacking access to adequate healthcare, recent innovations in the sector have flipped this conventional definition on its head and established a new balance of power moving forward. Rather than the formerly tried-and-true path of consumers from emergent nations traveling to so-called First World countries to receive care, exorbitant pricing and lack of transparency from insurance providers have resulted in people from the United States, the United Kingdom, and other high-income countries to subsequently seek out treatment for their own health afflictions outside of their native borders and on a global scale.

Medical, Dental, and Wellness Tourism: A Postpandemic Perspective. Mary Schreiber Swenson, PhD, and Amit Bansal (Authors).

Though the term "medical tourism" has been cemented into popularity over the years due to its all-encompassing nature, the reference is not altogether accurate for the current status of the industry and the misnomer merely lingers as a remnant of the past (Horowitz, 2007). Taking into account the disposable income bracket of the majority of 21st-century consumers that seek care abroad, coupled with major advancements in emerging countries' medical care, industry experts have pivoted to rename this new era of medical tourism as "travel for treatment" as a more precise reflection of the present-day trends, while "medical tourism" continues to be used in reference to cases where patients couple their medical-based travel with recreational activity in their destination of choice. Likewise, domestic medical tourism—where US citizens stay in the United States and utilize the best of the country's nationwide resources—has become an increasingly important sector in the global travel for treatment industry.

Given the higher income threshold for travel for treatment consumers when compared to what we traditionally think of as medical tourism patients, the movement has been able to expand rapidly en masse over the past couple of decades. While Horowitz's (2007) figures estimate that some 750,000 Americans traveled outside of the nation's borders to receive foreign care in 2007, that same number more than doubled by 2017 when some 1.4 million US citizens participated in the travel for treatment industry (Dalen and Alpert, 2019).

Part of this rapid growth can be explained by the increased travelability of the world for middle- to upper class persons in the new millennium, causing countries across the world to expand their inherent infrastructures to accommodate the tourists arriving at their borders. Estimated to grow at a compounded 4.3 rate internationally per annum (barring the widespread travel halt brought on COVID-19 pandemic) according to data from 2018, international travel has only been bolstered higher by the growth of major commercial airports to some 1200-plus hubs worldwide by the end of that very same year (Airports Council International, 2019).

For members of the US population, a quick drive down south of the border into the neighboring country of Mexico has opened up a wealth of possibilities for Americans to find quality care outside of their nation's boundaries, while a train or car ride north opens up possibilities within Canada's burgeoning medical tourism sector (Purdy, n.d.). Considering an increase of American drivers on the road from 167 million in 1990 to approximately 222 million drivers in 2016 (Wagner, 2021), the high number of licensed US drivers has resulted in an increased opportunity population-wide to travel for treatment.

With the *how* of Americans traveling for treatment established, it's vital to explain *why*. One clear factor has been public perception of the US healthcare system; while 73% of the US adults said they had great confidence in the country's medical leaders when surveyed in 1966, only 34% of Americans adhered to this same faith in the nation's health professionals nearly half a century later in 2012 (Baker, 2020). This issue was only exacerbated throughout the global COVID-19 pandemic, as stresses to the medical industry caused 32% of US citizens to lower trust in their healthcare providers throughout the COVID-19 pandemic (Building Trust, 2021).

These statistics are similarly skewed in favor of the American white populis; when factoring input in from marginalized communities, including members of the US Latino, Black, Asian, and Native American populations, this lack of trust is only amplified, with four out of five of this surveyed group of peoples reportedly refusing to return to previous healthcare providers after they had lost faith in their efficacy and operations (Read et al., 2021). Subsequently, more than a third of participants in the same study revealed they actively avoid seeking care for their medical issues due to this mistrust, highlighting deep-seated issues in the healthcare infrastructure of an increasingly diverse nation.

Insurance companies certainly have not helped the public perception of American healthcare, oftentimes presenting more challenges and difficulties to consumers than assisting with the provision of adequate treatment. According to a recent study by Bend Financial in the midst of the COVID-19 pandemic, more than 56% of surveyed US consumers were "completely lost when it comes to understanding health insurance" (Bend Financial, 2021), with another 65% unable to differentiate between insurance providers plan options of health savings accounts and flexible spending accounts.

Given that 65.8% of American adults are currently enrolled in private healthcare systems as opposed to public coverage options provided by the US government (Rosso, 2021), the American insurance conundrum has been inherently plagued by a lack of transparency for some decades now, only contributing to consumers' widespread confusion about pricing and treatment opportunities. Even public insurance choices like Medicare can come laden with uncertainty, as the US government's propensity to outsource administrative aspects to private companies contributes to inflated costs and prevents full consumer clarity on price structure, subsequently resulting in American adults paying approximately $111 more than necessary per month on their coverage plans (Bend Financial, 2021).

The problem, however, is systemic; the US healthcare system was designed specifically to financially benefit insurers and providers, leaving US consumers purposefully in the dark to accomplish their capitalistic goals. Since 1979, insurance companies have been protected by some iteration of the Uniform Trade Secrets Act across 48 states, allowing them to mainly keep their pricing structures secret from their customers—though not without controversy, yet few public interest exemptions have been granted on the matter (Gudiksen and Chang, 2019). As such, consumers cannot properly ascertain the actual cost of their treatment when making inquiries to their providers, preventing them from comparing competitive rates and picking the best option available to them, all while providers in the $1.151 trillion private healthcare market reap in the financial rewards in the meantime (Centers for Medicare and Medicaid Services, 2020).

This transparency issue then compounds back onto consumers, as choosing privatized healthcare options with exorbitant pricing can actually cause customers to receive less effective care and spend more on medications, despite already paying a high premium.

> From 2015 to 2020, patient out-of-pocket healthcare spending increased from $338 to $425 billion. It's been empirically established that higher out-of-pocket costs—for example, deductibles, co-payments for physician visits, cost sharing for prescription drugs—can lead to poorer adherence to treatment. The rise in patient cost sharing for pharmaceuticals, especially in the form of co-insurance as a percentage of the list price of medically necessary products, such as insulin, has been well-documented. (Cohen, 2020)

The United States' tumultuous and opaque healthcare system stands in stark contrast to the public coverage options provided by other First World nations, despite US consumers spending nearly double the average amount on healthcare in comparison to their peers (Schneider and Sarnak, 2017). In fact, most major nations with well-funded socialized healthcare, including the United Kingdom, Australia, Sweden, the Netherlands, Norway, and New Zealand, consistently outperformed the United States on nearly every single ranking metric established by the Commonwealth Fund. Among the 11 countries studied, the United States' coverage system ranked last in ease of access, equity, and healthcare outcomes, runner-up in administrative efficiency, and right in the middle of the pack on care process, begging the question of why American consumers are so reliant on a broken system—a pressing matter many are now seeking to resolve through travel for treatment.

While many US adults indeed possess insurance coverage, though the efficacy of said plans is highly debatable, a large number of Americans are barred from receiving adequate medical care due to their own lack of coverage and the United States' lack of universal healthcare. An estimated 50 million people, or 16% of the US population (Chemweno, 2021), have no form of healthcare coverage whatsoever, either unable to afford plans themselves or unprovided for by their employer, forcing them to instead pay exorbitant out-of-pocket prices for care.

Unfortunately for those living paycheck-to-paycheck, large American corporations in control of retailers and restaurants regularly schedule their employees' right below 40 h of work per week and reclassify workers as part-time, allowing them to circumvent the Obama-era health reforms that require employers to provide their full-time workers with health insurance and exacerbating the nationwide coverage issue (McVeigh, 2013). With multi-million dollar companies valuing saving pennies over the wellbeing of their minimum wage employees through this legal-yet-immoral bypassing of workers' benefit protection mandates, American adults already struggling with income are then left without assistance on their medical bills, trapping them in a vicious cycle of economic hardship that's becoming increasingly hard to break. This goes on to affect the American workforce as a whole, contributing to the US population's lower life expectancy and higher rate of chronic illnesses when compared to rival first-world nations.

1.2 DENTAL

Even American workers fortunate enough to receive insurance coverage from their employers are still required to pay out-of-pocket for certain aspects of healthcare like dental, as a significant number of health plans do not offer care for teeth as a part of their coverage packages. As a result, some 28% of surveyed working US adults reportedly do not have any type of dental insurance, with another 51% reportedly only visiting the dentist for serious issues surrounding their teeth. Post-retirement, the statistics grow even more grim; more than half of surveyed participants say they do not know how they will obtain or afford dental insurance after age 65 (Reinberg, 2017).

This lack of accessible care is directly reflected in the health status of the US population's teeth; according to the Centers for Disease Control, some 25.9% of American adults aged between 20 and 44 have untreated tooth decay, an issue that can almost only be remedied by the preventative care of professional dentists and numbers on par with the percentage of US adults uninsured for dental (Centers for Disease Control and Prevention, 2021). Given the consistency of these figures throughout, and the direct correlation between visiting the dentist regularly and the prevention of long-term tooth decay, it is easy to surmise that offering attainable and affordable dental care is a necessity to significantly improve the overall dental health of American consumers.

Unfortunately, this health-first approach has not been adopted by dental insurance providers nor holistically by the US government, leaving consumers of all incomes searching for a solution without an answer within their native borders. Enter traveling for treatment, with two of the world's most in-demand countries for dental medical tourism—Mexico and Costa Rica—no more than a drive or plane ride away for most American adults.

Standing in stark contrast to the United States' approach, Mexico's out-of-pocket prices have remained at a consistent fraction of the cost of American care, without sacrificing quality of care. Even Mexico's total number of dentists is nearly double that of the United States as of figures from 2019, featuring 120 dentists per 100,000 people as opposed to America's 61 dentists per 100,000 citizens and inherently offering two times the access to trained dental professionals (Elflein, 2021). With a wealth of qualified Mexican dentists to choose from that have no insurance barrier to begin treatment, Mexico's dental cost-efficient pricing structure has become an incredibly enticing feature for Americans looking for affordable care, resulting in more than 6 million US citizens traveling to the neighboring country for dental care on average per annum (Mendoza, 2021).

This demand is only bolstered when directly comparing the average rates for common treatments between the two nations; for common yet necessary procedures like root canals, US dentists charge between $1000 and $1500 on average for the treatment, while Mexican dental professionals offer the same service at up to 80% less in cost (Rivera, 2021). This extends into cosmetic treatments as well, as a set of high-end porcelain veneers can carry a price tag of at least $2000 per tooth in the United States, while a single porcelain tooth of the same quality is only a fraction of the cost in Mexico at an average price of $330 (Newman, 2020).

The demand for US consumers seeking care in Mexico is clear, as the 5000-person-strong town of Las Algodones, Mexico—located just south of the Arizona border and the Colorado river—is called home by 350 dentists, a strong 7% of the municipality's population. With such a high number of dental professionals available to work at a nominal cost in comparison to their US counterparts, Americans have continually flocked to the small town in spades, with some figures estimating around 3000 US citizens cross the border on a daily basis to receive dental care.

To Healthcare.com co-founder Jeff Smedsrud, recent consumer trends toward high-quality Mexican care over American options is anticipated to continue moving forward.

> It's likely less costly, most of the time the quality of care is going to be as good and it's a way to stretch your budget. There are significant savings opportunities, 20 to 50 to 70 percent, it's absolutely to the benefit of consumers to consider traveling to a different country for different kinds of procedures. (Rivera, 2021)

Americans have similarly looked across the pond for dental treatment opportunities, with Eastern European countries like Poland and Slovakia offering competitive pricing for quality care. Europe's Schengen Area has likewise enticed US citizens to travel abroad for dental care, as the 10-country group hosts three of the globe's top dental tourism destinations—the Czech Republic, Spain, and Hungary—all accessible through the 90 days of visa-free access to the region made available to US passport holders (Newman, 2020).

1.3 INTENSIVE MEDICAL TREATMENTS

While dental procedures remain a perennially popular choice in travel for treatment, the medical tourism industry similarly boasts robust sectors globally dedicated to chronic illnesses and serious surgeries that can become unaffordable in the United States. In the case of cancer, some 63% of surveyed cancer patients and their close relatives reported financial difficulties surrounding their ongoing cancer treatments, with the price tag of necessary monthly pharmaceuticals often eclipsing the $100,000 threshold with insurance involved (Selby, n.d.).

As one of the top killers of the American population and second only to heart disease, the United States spent $150.8 billion on cancer care in 2018 in an effort to combat the illness. Still, exorbitant out-of-pocket prices have forced the hand of low-income workers and uninsured persons unable to afford treatment to look abroad for their care, while members of other US tax brackets similarly dissuaded by rising price tags travel for cancer treatment in tandem to destinations renowned for their expertise on the subject, like Israel and India (Rastogi et al., 2020).

Aforementioned commonplace heart issues have too spurred a rise in medical tourism for American patients, particularly when considering the most common cardiothoracic surgical procedure in the United States, the coronary artery bypass graft procedure, accounts for $6.5 billion in health-care costs across the country on an annual basis as one of the country's most expensive surgeries available (Mori and Khera, 2020). As such, consumers of all economic backgrounds have sought care on foreign soil, looking to receive comparable surgical results at a portion of US prices.

> The cost savings can be considerable. The price of a coronary artery bypass graft procedure (CABG) for example, may be as little as $5,000, including travel expenses, compared with a cost of $33,000 to $63,000 in [the United States]. (Jacobs et al., 2013)

Oftentimes, cardiothoracic surgeries can need near-immediate attention or patients could face potentially irreversible consequences to their health or ultimately lose their lives. Despite this, the United States has remained notorious for its long waitlists for heart surgeries and other major organ operations, leaving a negative impact on cardiac surgical patients that can be remedied through medical tourism (Eberechukwu, 2020). Looking abroad can allow consumers to receive the care they need at a much faster rate, all while saving money in the process, making travel for treatment an attractive option for many Americans.

US citizens looking to partake in cosmetic elective surgeries like rhinoplasties and liposuctions can likewise benefit from participation in the medical tourism sector, with countries like India, South Africa, and Costa Rica reportedly becoming increasingly popular destinations for plastic surgery and cosmetic procedures (Franzblau and Chung, 2013).

1.4 EVALUATING OPTIONS

Another promising aspect of travel for treatment has been the increased opportunities for potential patients to pursue their various global options for care, benefitting from a second opinion on an international scale. Rather than being constricted to medical advice given within the confines of the teachings of US medical schools, increasing queries and scope onto a worldwide basis allows American consumers to consider adopting treatments using emerging technology or state-of-the-art methodology not yet approved for use in the United States.

> ...There is an argument that servicing the needs of foreign patients may broaden the case-mix for staff, or may increase throughput to enable them to become more skilled; it might open up the door to secondments to overseas facilities which, provided migration is temporary, may lead to enhancement of human capital; it may provide increased quality through ensuring compliance with (higher) international standards for care (as alluded to above); and it may promote a culture of personal development in skills and technologies available to treat patients generally, which local patients will of course benefit from. (Lunt et al., 2022)

Some countries outside simply have more equipment and machinery available than the United States, allowing for a more comprehensive view of a patients' medical diagnosis; while the United States has 42 computed tomography (CT) scanners per million inhabitants, a tech-forward country like Japan possesses nearly triple the amount of the same machinery, offering 111 CT scanners for every million residents as per figures from 2020 (OECD, 2017). On the same note, Japan likewise possesses three of the world's six tumor-treating carbon ion radiotherapy facilities; the United States hosts zero within its borders as of 2021 (Mayo Clinic, 2020).

Considering the healthcare industry's continuous evolution alongside 21st-century technology, the value of a second of a second opinion has become increasingly recommended by medical professionals (Katella, 2020), particularly given varying degrees of equipment availability on a country-by-country basis. This allows patients to adequately understand global pricing diversities, potentially obtain a correct diagnosis, and target the exact region of expertise for necessary treatments. When utilized correctly, travel for treatment helps provide American consumers quality and cost-efficient care that would otherwise be inaccessible to them within the United States, similarly offering patients innovative and modern international medical solutions they would otherwise never have the opportunity to explore.

1.5 CENTERS OF EXCELLENCE

While an entire new world of healthcare opportunities has opened up by the expansion of the medical tourism industry, the tidal wave of new information can be overwhelming for patients in search of a cure for their wellness problems. As such, Centers of Excellence (CoE) have become increasingly dependable resources for Americans and other international consumers traveling for treatment. Composed of specialized programs within preestablished medical facilities, CoEs heavily emphasize quality standards in operations while simultaneously retaining consistently high concentrations of expertise in the centers' related fields. Given CoEs meticulously organized infrastructure and attentiveness to specific medical conditions, the programs' efficiency have routinely enticed in a specialized patient pool generally unseen by standard hospitals, though currently, there remains no formalized accreditation process to determine what constitutes a COE, relying instead on facilities' self-proclamation of expertise and capabilities (Elrod and Fortenberry, 2017).

> The very tight, centralized organization designs used in properly assembled centers of excellence foster many of the benefits associated with this particular delivery model. Patient convenience is maximized by consolidating previously dispersed care components, improving satisfaction. The formally-structured, concentrated focus on particular medical conditions permits specialization, affording tailored care experiences which feature customized environments and expert personnel. The centralized structure also pools resources which otherwise would be distributed more broadly, with this fostering efficiency and effectiveness. (Elrod and Fortenberry, 2017)

This is particularly pertinent in the case of travel for treatment and medical tourism, as a patient's flexibility to visit CoEs in countries outside of the United States would potentially allow them to find a program tailor-made for their individual medical conditions, boosting treatments' overall efficacy in turn. While the cost benefits of healthcare abroad have already been determined, CoEs help ascertain patients' access to international care of comparable or heightened quality to that of the United States. Still, there has not yet been a standardized data-based criterion implemented to position one CoE as more qualified than the next, leaving the programs on an even playing field and helping foster an equitably competitive international marketplace in the process.

1.6 THE STATUS OF SOCIALIZED MEDICINE IN THE UNITED STATES

This same fair market approach is hardly replicated within the United States; while CoE have similarly been constructed within its own borders, the domination of private insurers in the healthcare sector has prevented many lower income and uninsured American people from benefiting from CoEs niche expertise, an experience that's likewise economically barred many potential medical tourists to the United States from further exploring their options—though the United States is anticipated to grow its own internal travel for treatment sector by 20% per annum (Beladi, 2017).

Still, the status quo of privatized coverage's near-full command of the US healthcare system has been decreasingly well received by American consumers over the years, as modern digitization efforts and the ubiquity of the internet have led to a better informed population in turn (Cline, 2001). As of 2021, some 56% of Americans surveyed in a Gallup poll said that the nation's healthcare structure should be the full responsibility of the US government. Though these figures were likely bolstered by the arrival of the COVID-19 pandemic and the subsequent health-related issues the illness brought with it, American consumers' public opinion has been steadily swaying toward increased government intervention since 2014 (Gallup, 2022).

While adopting a universalized healthcare system akin to the United Kingdom's would immediately benefit the wellness of America's low-income families and individuals, this same shift in methodology could similarly help reduce inequality in the United States on a societal scale (Marmot, 2012), leading to higher rates of population satisfaction and theoretically amplifying the ability for more consumers to participate in the American economy. This would then likewise contribute to revenue increases for top-earners' businesses in turn, positioning socialized healthcare's implementation in the United States as anticipatorily favorable across all of the country's social classes. Still, this potential future is still far off, as privatized healthcare continues to reign supreme and multiple state systems have begun implementing legislation crafted specifically to deter US consumers from seeking treatments outside of the United States. However, as with most state-governance and private sector matters, there is variation; some states have gone on to actively encourage their residents to find more cost-efficient medical care abroad, while some choice insurance providers have begun taking travel for treatment into account in their coverage packages.

> The United States health insurance is complicated, and there have been several difficulties expanding because there are state and federal regulations and legislation and even politics involved. For example, price incentives are not allowed in the state of Texas to medical tourists. The law was put in place to stop doctors from limiting the request of medical tourism in Mexico due to the proximity. Despite that, some states have given the

> go-ahead for employees to travel overseas for medical tourism, such as California, where medical tourism is strongly encouraged. Furthermore, some private health insurance companies such as Seven Corners have drafted out packages like Bordercross Worldwide (plan) to cover medical treatment for their clients abroad medical tourism. (Eberechukwu, 2020)

In this same vein, self-insured private corporations and Fortune 500 companies can analogously take advantage of the growing boons of travel for treatment, now possessing the unique opportunity to craft bespoke care for their roster of employees. Considering companies regularly pay for private coverage for their staffers priced in line with the United States' documented sky-high costs, crafting a benefit plan around medical tourism as an asset could allow businesses to significantly cut down on their payroll healthcare expenditures, while still giving employees full access to quality care in. A self-funded approach such as this likewise has organizations assume the risk of travel for treatment rather than individuals, allowing businesses to curate plans based on need and reduce their tax liability, while workers can acquire the care they need sans personal risk (Medical Tourism Magazine, n.d.).

1.7 THE ADVENT OF TELEHEALTH

While telehealth was brought to the forefront of available healthcare services for US consumers during the COVID-19 pandemic, an expansion particularly spurred on by lack of access to in-person care due lockdown measures and social distancing regulations, the sector has been contributing to the medical tourism industry for some time now. Though many, particularly in the United States, have been hesitant to classify telehealth services as "travel for treatment," the concept's integral ethos remains, as consumers continue to seek care outside of the standardized confines of the US healthcare system—just from the comfort of their computers rather than in a foreign country. This allows for enhanced access to specialists and improved convenience, allowing professionals from different locales to work together with one another on treating specific medical conditions while also offering more affordable rates than traditional brick-and-mortar care (Ferreira and Rosales, 2020).

Though telehealth is becoming an increasingly important asset to US consumers seeking quality and affordable care, particularly after COVID-19 boosted telehealth to levels 38 times higher than prepandemic numbers (Schwietzer, 2022), the virtual practice was utilized by other foreign countries long before the pandemic's arrival. With the burgeoning industry expected to continue its upward growth trajectory, a trend particularly bolstered by 91% of large US employers now offerings virtual mental healthcare as part of their employee coverage practices (Ferreira and Rosales, 2020), telehealth is anticipated to become more and more commonplace once the COVID-19 pandemic has passed, as American consumers' newly discovered access to easily acquired and quality care is expected to lead to continued surges in demand for the virtual services moving forward.

KEYWORDS

- **medical tourism**
- **travel for treatment**
- **healthcare**
- **telehealth**
- **second opinion**
- **centers of excellence**

REFERENCES

Airports Council International. *Preliminary World Airport Traffic Rankings Released*; Airports Council International, Mar. 13, 2019. https://aci.aero/2019/03/13/preliminary-world-airport-traffic-rankings-released/ (accessed Jan. 19, 2022).

Baker, D. W. Trust in Health Care in the Time of COVID-19. *JAMA* **2020,** *324* (23), 2373. https://doi.org/10.1001/jama.2020.23343

Beladi, H.; et al. Does Medical Tourism Promote Economic Growth? A Cross-country Analysis. *J. Travel Res.* **2017,** 58 (1), 121–135. https://doi.org/10.1177/0047287517735909.

Bend Financial. *More than Half of Americans Confused by Health Insurance, Including HSAs*; Bend Financial, 2021. https://www.bendhsa.com/newsroom/more-than-half-of-americans-confused-by-health-insurance-including-hsas

Building Trust. *Public & Physician Trust in the U.S. Health Care System*; Building Trust. https://buildingtrust.org/public-physician-trust-in-the-u-s-health-care-system/

Centers for Disease Control and Prevention. *FastStats—Oral and Dental Health*; Centers for Disease Control and Prevention, Aug. 3, 2021. https://www.cdc.gov/nchs/fastats/dental.htm

Centers for Medicare and Medicaid Services. *National Health Expenditure Data Fact Sheet*; Centers for Medicare and Medicaid Services, 2020. https://www.cms.gov/Research-Statistics-Data-and-Systems/Statistics-Trends-and-Reports/NationalHealthExpendData/NHE-Fact-Sheet.

Chemweno, J. The U.S. Healthcare System Is Broken: A National Perspective. In *Managed Healthcare Executive*, July 27, 2021. https://www.managedhealthcareexecutive.com/view/the-u-s-healthcare-system-is-broken-a-national-perspective

Cline, R. J. Consumer Health Information Seeking on the Internet: The State of the Art. *Health Educ. Res.* **2001,** *16* (6), 671–692. https://doi.org/10.1093/her/16.6.671

Cohen, J. U.S. Healthcare Markets Lack Transparency; Stakeholders Want to Keep It That Way. *Forbes Magazine*, Dec. 7, 2020. https://www.forbes.com/sites/joshuacohen/2020/12/07/us-healthcare-markets-lack-transparency-stakeholders-want-to-keep-it-that-way/?sh=6a3111ec62b2

Dalen, J. E.; Alpert, J. S. Medical Tourists: Incoming and Outgoing. *Am. J. Med.* **2019,** *132* (1), 9–10. https://doi.org/10.1016/j.amjmed.2018.06.022

Eberechukwu, K. *Medical Tourism: The Impact of Cost and Wait Time in Relation to Cardiac Surgeries among Individuals in the United States*; Marshall University, Marshall Digital Scholar, 2020; pp 2–11.

Elflein, J. *Oral Health and Dental Care in the U.S.*; Statista, Aug. 2, 2021. https://www.statista.com/topics/3944/oral-health-and-dental-care-in-the-us/#dossierKeyfigures

Elrod, J. K.; Fortenberry, J. L. Centers of Excellence in Healthcare Institutions: What They Are and How to Assemble Them. *BMC Health Serv. Res.* **2017,** 17 (S1). https://doi.org/10.1186/s12913-017-2340-y

Ferreira, W. F.; Rosales, A. International Telemedicine: A Global Regulatory Challenge. *Lexology*; Hogan Lovells, Mar. 9, 2020. https://www.lexology.com/library/detail.aspx?g=f2d9946b-e5c3-43f5-b813-9528e23afbda

Franzblau, L. E.; Chung, K. C. Impact of Medical Tourism on Cosmetic Surgery in the United States. *Plast. Reconstruct. Surg. Glob. Open* **2013,** *1* (7). https://doi.org/10.1097/gox.0000000000000003

Gallup. *Healthcare System*; Gallup.com, Gallup, Jan. 26, 2022, https://news.gallup.com/poll/4708/healthcare-system.aspx

Gudiksen, K. L.; Chang, S. M. California Health Care Foundation. *The Secret of Health Care Prices: Why Transparency Is in the Public Interest*, 2019. https://www.chcf.org/wp-content/uploads/2019/06/SecretHealthCarePrices.pdf (accessed Jan. 19, 2022).

Horowitz, M. D.; et al. Medical Tourism: Globalization of the Healthcare Marketplace. *MedGenMed: Medscape General Medicine*; Medscape, Nov. 13, 2007. https://www.ncbi.nlm.nih.gov/pmc/articles/PMC2234298/

Jacobs, J. P.; et al. Surgical Tourism: The Role of Cardiothoracic Surgery Societies in Evaluating International Surgery Centers. *Ann. Thorac. Surg.* **2013,** *96* (1), 8–14. https://doi.org/10.1016/j.athoracsur.2013.02.058

Katella, K. Can a Second Opinion Make a Difference? *Yale Medicine*, Jan. 15, 2020. https://www.yalemedicine.org/news/second-opinions

Lunt, N.; et al. Image Result for OECD Image Result for OECD More Images Organisation for Economic Co-Operation and Development, *Medical Tourism: Treatments, Markets and Health System Implications: A Scoping Review*. 2022. https://www.oecd.org/els/health-systems/48723982.pdf (accessed Jan. 27, 2022).

Marmot, M. Why Should the Rich Care about the Health of the Poor? *Can. Med. Assoc. J.* **2012,** *184* (11), 1231–1232. https://doi.org/10.1503/cmaj.121088

Mayo Clinic. *Mayo Clinic Plans First Carbon Ion Therapy Facility in North America*. Mayo Clinic. Mar. 16, 2020. https://www.mayo.edu/research/forefront/mayo-clinic-plans-first-carbon-ion-therapy-facility-in-north-america

McVeigh, K. US Employers Slashing Worker Hours to Avoid Obamacare Insurance Mandate. *The Guardian*; Guardian News and Media, Sept. 30, 2013. https://www.theguardian.com/world/2013/sep/30/us-employers-slash-hours-avoid-obamacare

Medical Tourism Magazine. *Infusing Medical Tourism into Corporate America*; Medicaltourism.com, Medical Tourism Magazine, n.d. https://www.magazine.medicaltourism.com/article/infusing-medical-tourism-into-corporate-america

Mendoza, J. *Medical Tourists in Mexico 2020*; Statista, July 2, 2021. https://www.statista.com/statistics/982297/mexico-tourists-medical-purposes/#:~:text=Mexico%3A%20number%20of%20medical%20tourists%202020&text=In%20March%202020%2C%20around%20728,same%20month%20a%20year%20earlier

Mori, M.; Khera, R. Pumping the Breaks on Health Care Costs of Cardiac Surgery by Focusing on Postacute Care Spending. *Circ.: Cardiovasc. Quality Outcomes* **2020,** *13* (11). https://doi.org/10.1161/circoutcomes.120.007253

Newman, E. Worth the Trip? A Look at Dental Tourism. Ontario Academy of General Dentistry, Academy of General Dentistry, Jan. 20, 2020. https://www.agd.org/constituent/news/2020/01/20/worth-the-trip-a-look-at-dental-tourism

OECD. *Computed Tomography (CT) Scanners*; OECD, 2017. https://doi.org/10.1787/bedece12-en

Purdy, L. *Deloitte Canada, Evolving Medical Tourism in Canada*. n.d. https://www2.deloitte.com/content/dam/Deloitte/ca/Documents/life-sciences-health-care/ca-en-health-care-life-sciences-evolving-medical-tourism-exploring-a-new-frontier.pdf (accessed Jan. 18, 2022).

Rastogi, K.; et al. Analysis of Medical Tourism of Cancer Patients in Asia-Africa-Oceania Region and Financial Toxicity Due to Migration. *Int. J. Radiat. Oncol. Biol. Phys.* **2020,** *108* (3). https://doi.org/10.1016/j.ijrobp.2020.07.2494

Read, L.; et al. Rebuilding Trust in Health Care. *Deloitte Insights*; Deloitte, Aug. 15, 2021. https://www2.deloitte.com/us/en/insights/industry/health-care/trust-in-health-care-system.html

Reinberg, S. Can't Afford the Dentist? You're Not Alone. *CBS News*, CBS Interactive, Sept. 7, 2017. https://www.cbsnews.com/news/cant-afford-the-dentist-youre-not-alone/

Rivera, S. Study Finds Medical and Dental Care South of the Border Are Good Options for Americans. *BorderReport*, July 28, 2021. https://www.borderreport.com/health/study-finds-medical-and-dental-care-south-of-the-border-are-good-options-for-americans/

Rosso, R. J. Congressional Research Service, 2021. *U.S. Health Care Coverage and Spending*. https://sgp.fas.org/crs/misc/IF10830.pdf (accessed Jan. 19, 2022).

Schneider, E. C.; Sarnak, D. O. Mirror, Mirror 2017: International Comparison Reflects Flaws and Opportunities for Better U.S. Health Care. *Mirror, Mirror 2017: International Comparison Reflects Flaws and Opportunities for Better U.S. Health Care*; July 2017. https://interactives.commonwealthfund.org/2017/july/mirror-mirror/

Schwietzer, R. Council Post: Will the Telehealth Boom Continue after the Pandemic? *Forbes Magazine*, Jan. 7, 2022. https://www.forbes.com/sites/forbestechcouncil/2022/01/06/will-the-telehealth-boom-continue-after-the-pandemic/?sh=da9144f52719

Selby, K. *Americans Can't Keep Up with High Cost of Cancer Treatment*. Mesothelioma Center—Vital Services for Cancer Patients & Families, n.d. https://www.asbestos.com/featured-stories/high-cost-of-cancer-treatment/

Wagner, I. *Car Drivers—Statistics & Facts*; Statista, Apr. 15, 2021. https://www.statista.com/topics/1197/car-drivers/#dossierKeyfigures

CHAPTER 2

HOW THE COVID-19 PANDEMIC HAS AFFECTED THE TREATMENT SECTORS

ABSTRACT

Considering the medical tourism industry is built on the back of international travel, the border restrictions and national quarantine policies brought on by the COVID-19 pandemic changed the dynamics of the medical tourism industry as we know it. This chapter explores the shift between pre-pandemic and post-pandemic tourism numbers across the world, how consumer health trends subsequently adapted to the "new norm," and the accelerated rise of virtual healthcare in lieu of in-person opportunities.

Now that we have established the definition, the basis and the methodologies of traditional travel for treatment on an international scale, it's imperative to explore the impact of the worldwide COVID-19 pandemic on the health and wellness industry worldwide (International Trade Administration, n.d.). Between massive hits to the US population's overall health, exacerbated economic inequalities, and employer layoffs removing able workers' needed insurance options, the pandemic has made a number of disheartening situations commonplace among the American working class, with travel restrictions further preventing seeking care outside of the United States' borders. The same issues have affected international travelers in turn; pandemic-born measures preventing the travel of foreigners to the United States, which claims a major take of 28% of the total global wellness market, and other popular medical tourism countries, as well as simultaneous economic hardships, largely grinded the travel for tourism industry to halt on a worldwide scale (Global Wellness Institute, 2021).

Medical, Dental, and Wellness Tourism: A Postpandemic Perspective. Mary Schreiber Swenson, PhD, and Amit Bansal (Authors).

Still, while COVID-19 may have undeniably impacted the average consumer seeking medical care negatively, innovations made within the difficult time period have begun setting up global patients for increased accessibility moving forward. Between further forays into the telehealth space and an overall enhanced global focus on wellbeing and vitality, a number of pandemic-born inventions and evolutions to preestablished wellness spaces are setting the stage for an anticipatedly improved medical tourism sector upon the pandemic's close.

2.1 INTERNATIONAL TROUBLES

Given the heavy lockdown measures and social distancing restrictions implemented by many countries across the world, the pandemic has fundamentally changed the dynamics of travel, health, and wellness care and how patients navigate their ways through the system. In line with other infectious disease outbreaks, such as 2009's H1N1 and the Ebola scare of the early aughts, countries across the world began closing their borders to outside travelers after COVID-19 began making an international impact in March 2020 (Gwee et al., 2021).

While the ultimate long-term impacts of completely shutting down nations from global tourism and consumerism remains to be seen, respected health institutions like the World Health Organization have largely denounced the need for travel restrictions, finding the overall impact on COVID-19's transmission to be near negligible unless all carefully placed social distancing regulations are followed to the fullest extent. However, country after country—including the United States—began implementing international travel bans and border closures at the start of the pandemic, halting economic activity and consequently impacting their populations in the process.

As the origin region of the COVID-19 outbreak, countries across the Asian continent were some of the first in the world to impose travel restrictions on foreign tourists and implement new health policies among their citizens. Considering the emergence of the virus from Wuhan, China, the Chinese government immediately shut down any entrances or exits from the Hubei province in an attempt to curb COVID-19's spread within the nation. On a national scale, China took a zero-COVID policy and made the decision to close its borders to all foreign travelers, similarly requiring its population to stay at home. Meanwhile, the country itself experienced bans from dozens of other countries preventing Chinese citizens from traveling to these various locales, limiting the impact of China's 1.4 billion-strong population on the economies and medical tourism industries of international countries previously frequented by Chinese tourists (Zhang, 2021). In the case of South Korea, the nation's government chose to put in place robust testing measures and intensive tracing measures, which allow its population to largely move freely throughout the country if not infected, standing in contrast to China's stringent approach (Wang et al., 2022). For foreign tourists, Korea implemented a relatively lax policy when compared to its nearby neighbor, allowing international travelers to enter the country after a 14-day mandatory quarantine and subsequently reducing COVID-19's overall impact on the South Korean economy when compared to other developed nations (Lim et al., 2021).

While responses to the first wave across Europe varied country-by-country, nations hit hard at the onset of COVID-19 like Italy provided a roadmap for other countries on the continent to prepare for the virus's impending hit within their borders, causing many—such as Germany, the Netherlands, and Hungary—to put moderate lockdown measures in place and restrict outside travelers in an attempt to slow COVID-19's spread. Others, such as Sweden and Norway, avoided going into full lockdown by implementing stay-at-home orders within their populations, though this still resulted in limited economic and tourist activity at the time (Plümper and Neumayer, 2020).

In the Middle East, stringency levels to lockdown varied according to the political powers controlling the country, which already stand in stark contrast to the largely democratic leanings of European countries. For Jordan, which is ruled by an authoritarian regime, the country chose to go for full-out closure despite a relatively low-case count on a global scale. In contrast, Turkey, which is comparably more democratic than its region counterparts, declined to implement any legal lockdown measures throughout the pandemic until 2021, a decision perhaps bolstered by its possession of the world's largest number of hospital beds, though it did suspend elective surgeries for a brief two-month period during the first wave (Gjerløw and Jensehaugen, 2020; Arslan et al., n.d.). This left the door open for travel for treatment opportunities for international tourists into Turkey post-April 2020; however, given the massive saturation of COVID-19 patients in the country's hospital beds, its respective openness compared to its Middle Eastern peers continued to limit the rate of elective surgeries in the country.

Compared to the diversified approaches taken by other regions, countries across the African continent banned together under the Africa Task Force for Coronavirus (AFCOR) to tackle the problem holistically. Under the AFCOR's guidance, the majority of African countries proceeded to ban travelers from outside countries with strong ongoing outbreaks, which was soon followed by local travel restrictions and social distancing measures. While these policies similarly hurt the collective African economy and limited medical tourism to the continent, Africa's collaborative effort and commitment to working together allowed it to make it through the first wave of the COVID-19 pandemic with an incredibly low death rate, particularly when put into comparison to the mortality rate among the populations of Europe and North America (Kuehn, 2021).

Within the North American market, these very border shutdowns, stay-at-home orders and lockdown measures grinded the US patients' opportunities for medical tourism to a halt out of sheer accessibility. As we explored previously, Mexico has long positioned itself as a cost-efficient and locationally favorable alternative to the American healthcare system, particularly among dental patients. However, with the United States closing its border by land with Mexico for twenty months throughout the COVID-19 pandemic, US consumers previously capable of seeking close-proximity care via a quick drive south were forced to largely scrap their travel for treatment plans (Diaz and Gonzalez, 2022). While Americans were allowed to enter Mexico by air throughout the same 20-month time frame, the cost factor of purchasing plane tickets and other travel arrangements naturally priced out many US medical tourists from seeking out further care, forcing them to instead rely on expensive US healthcare or wait with their ailments until travel restrictions eased.

The United States' northern neighbor of Canada, which had begun gaining global repute for its own appeal toward international medical tourists prior to the COVID-19 pandemic, similarly shut itself down to all but essential travel (Vieira and Mackrael, 2022). This expanded into closing down the US–Canadian border for more than a year of the pandemic, restricting both Canadians from entering the United States and Americans from entering Canada at the moment, and severely limiting both nationalities subsequent access to nearby foreign medical care during one of the most health-centric eras of the 21st century.

This in turn had repeatedly negative consequences on the service industry as a whole, including throughout the medical, dental, and wellness sectors, as widespread difficulties of businesses unable to financially compete in an inactive consumer marketplace resulted in more than 200,000 companies shutting down in the United States within the first year of the pandemic

(Simon, 2021). Of course, these massive closures compounded back onto the US working population; approximately 9.6 million laborers lost their employment status due to the effects of COVID-19—7 million more jobs lost than in the entirety of the European Union during the same time frame and the largest decline in employment in documented US history (Bennett, 2021). These layoffs and closures were recorded to disproportionately affect the service sector, hurting industries like hospitality and food services, and going on to negatively affect many deemed nonessential activities in the medical, dental, and wellness industries (US Bureau of Labor Statistics, n.d.).

> Businesses where social distancing was impractical, such as airlines, hotels, restaurants, and theaters, furloughed or let go a much higher percentage of their workforce than businesses whose workers could shift to remote work. For example, employment in food preparation and serving-related occupations fell by 48.8% between February and April 2020, while the number of people engaged in personal care and service occupations contracted 44.2%. (Buckley et al., 2022)

Even for Americans who retained their jobs throughout the pandemic, major roadblocks in the global supply chain made for a difficult era of US consumerism. With widely varying policies on a global scale, countries who fought to keep economic activity moving internally were set back by their peers' approaches, with port shutdowns, temporary trade restrictions, and labor shortages all contributing to the magnification of the issue.

> …the global cost of the pandemic depends foremost on the number of affected countries, and then on the required duration of lockdown policies; by contrast, the strictness of these policies is comparatively less important. The spatial extent of the pandemic is the most important driver of the global cost. If only China had been affected, our results suggest that the global supply-chain effects (measured by value added) would have been 3.5% of global gross domestic product. With the spread to highly developed western countries and the containment measures placed in Europe and the United States, we found that the global supply-chain effects increase almost fourfold to 12.6%. (Guan et al., 2020)

The supply chain issue has been well documented in its effects on the medical industry, particularly when taking into account the mad-dash toward acquiring personal protective equipment (PPE). While these accessories were once almost exclusively used by medical professionals, equipment like face masks and shields quickly became a necessity for the general public, prompting hoarding issues, price gouging, and a subsequent lack of available stock at the beginning of the pandemic. With much of the PPE held fast by enterprises looking to make a quick buck of the classic supply-and-demand model, health professionals working in facilities tending to COVID-19 patients and other ailments were left ill-equipped to fulfill their duties (World Health Organization, 2020).

Beyond the clear need for masks and gloves within hospitals, other industries like the dental, wellness, and aesthetic treatment sectors that heavily rely on regular PPE use to follow sanitary regulations were left hanging out to dry upon the COVID-19 pandemic's onset. Without the proper protection available, elective cosmetics administrators and dentists alike were left with the opportunity to continue their work at full capacity, finding their success limited by both aforementioned social-distancing restrictions and PPE supply problems.

> From March through August 2020, weekly visits to dental offices were 33% lower, on average, than in 2019. Weekly visits were 34% lower, on average, in counties with the highest COVID-19 rates.... Although rates of weekly visits rebounded substantially by June 2020, rates remain about 20% lower than the prior year as of August 2020. (Kranz et al., 2021)

These widespread supply chain issues ultimately trickled down beyond businesses to affect regular consumers, with many retail stores unable to function without proper stock of products temporarily shuttering their doors and in turn preventing sales from being made. Though e-commerce became an increasingly viable alternative for purchases throughout the pandemic, sales for wellness-related products dipped heavily, including a 32% decline in retail sales in the cosmetics industry at the beginning of the COVID-19 pandemic (Bjerke et al., 2020).

2.2 CONSUMER TRENDS AMID COVID-19

Between the supply hurdles, extensive layoffs, and business closures brought on by the pandemic, global consumer spending went on to drop by 4.2% from 2019 to 2020 according to data from Euromonitor International (n.d.). With millions of unemployed Americans unable to receive income from their lost professional careers, many struggling US consumers declined to spend their money on recreational items and services, instead diverting their funds to essential items and bolstering savings, a trend similarly reflected on an international scale.

Naturally, this low consumer demand had consequences on areas where consumers deem their purchases as "discretionary" or as "luxuries," affecting previously popular areas of medical tourism such as elective surgery, elective dental care, plastic surgery, and other elective medical procedures. According to data from the US National Library of Medicine National Institutes of Health, "there was a 91% reduction from baseline in the number of elective surgeries performed" in the United States at the onset of the pandemic between January 1 and September 30, 2020, a decrease brought on by both the need for hospital beds for COVID-19 patients that can similarly be attributed to changes in consumer spending trends (Prasad et al., 2021).

Within the dental care sector, guidance from the American Dental Association and decisions from state governments caused most dental offices to pause all elective procedures and only take on patients in need of emergency care for nearly three months, leaving only 5% of preestablished US offices open at the time and expert predictions of a "37.4 percent reduction in US dental spending in 2020 compared to pre-pandemic projections" (Nasseh and Vujicic, 2020). Though dental procedures have rebounded close to prepandemic levels in the months since the pandemic's beginning, the economic impact on the US dental industry still lingers.

Even the supply chain issues caused consumers to reassess their spending habits in the long term, with available stock turning into a primary asset for modern retailers. Statistics from December 2021 show that "over 60 percent of US consumers have experienced out-of-stock items in the last three months, and when this happened, only 13 percent waited for the item to come back in stock versus the 39 percent who switched brand or products and the 32 percent who switched retailers" (McKinsey & Company, 2021a).

However, while the average consumers' spending power was altogether limited by COVID-19, many high-income individuals were able to compound on their wealth at this time of economic upheaval, widening the wealth gap between the rich and poor on both a national and global basis. While the government's $5.4 trillion expenditure on economic stimulus and other support programs helped give lower income individuals a firmer ground to stand on at the start of the pandemic, subsequent declines in government support that temporarily mitigated loss have exacerbated the economic difficulties of the low-to-middle classes that were already disproportionately affected by layoffs (Buckley et al., 2022).

This inequality issue is best surmised by the soaring wealth of billionaires throughout the pandemic. With many average Americans spending their stimulus money on e-commerce marketplaces owned by the world's hyperwealthy like Amazon's Jeff Bezos and Walmart's Walton family, money spent on essentials went straight out consumers' bank accounts and right into the pockets of the world's richest individuals. In 2020 alone, global billionaires enjoyed a cumulative increase in networth of $3.6 trillion, while the pandemic pushed another 100 million people across the map into abject poverty—a stark contrast revealing just how differently the income classes experienced the COVID-19 pandemic (Luhby, 2021).

As such, while some wealthy consumers in high-income countries may not have changed their spending habits on health and wellness minutely throughout the pandemic, many middle- and low-income consumers had neither the time nor money to spend on wellness, medical, and dental care during the same time period.

In a similar vein, layoffs during the COVID-19 pandemic showed a clear disparity between men and women in the American workplace, with women both losing more jobs and taking a longer time for their employment market to rebound than their male counterparts. Based on 2020 data from the American Society of Plastic Surgeons (2020), women make up 92% of all elective plastic surgery and cosmetic procedures performed in the United States compared to men's 8% share; given the gender's pandemic employment issues, it's no surprise that the rate of elective surgeries dipped during this time period considering its most loyal demographic fiscal roadblocks.

> According to seasonally adjusted employment data from the US Bureau of Labor Statistics, female employment fell by 17.9% between February and April 2020, while male employment declined by 14.3%. And even though employment for both men and women has been recovering steadily, employment among women as of September 2021 was still 3.6% lower than pre-pandemic levels—worse than the 2.8% decline among their male counterparts. Consequently, the share of women in total employment has gone down marginally to 47.0% during this period. (Buckley et al., 2022)

Of course, given the capitalist foundations of the United States, consumers were still required to participate in the American economy throughout the pandemic in spite of any personal financial woes. However, given the new priorities brought on by the time's health-conscious climate and previously explored fiscal limitations, consumers began assessing exactly where they were putting their money, opening up a new door of pandemic-born industry evolutions and brand new businesses catering directly to this new consumer culture.

Considering the widespread closure of brick-and-mortar retail stores due to aforementioned social distancing regulations and lockdown measures, global shoppers began gravitating toward the online marketplace in droves, with international sector revenues increasing by 19% and 22% respectively in 2020 and 2021—an expansion more than double the prepandemic anticipated growth estimates of 9% and 12% (International Trade Administration, n.d.).

According to the US Census Bureau, American consumers alone spent $211.5 billion on e-commerce throughout 2020's second quarter, a marked uptick of 31.8% from the first quarter. Throughout this same period, e-commerce grew to account for 16.1% of all sales in the United States, a 4.3% increase from the previous financial quarter, a trend that's expected to endure in the years moving forward from the pandemic (JP Morgan, 2020). This too was reflected on an international scale, boosting the global e-commerce industry by 3% in 2020 from the year previous (United Nations Conference on Trade and Development, 2021).

Forging forward in the new-normal marketplace, "consumers across the globe have responded to the crisis and its associated disruption to normal consumer behaviors by trying different shopping behaviors and expressing a high intent (65% or more) to incorporate these behaviors going forward" (McKinsey & Company, 2020). This trend has likely been exacerbated by COVID's economic difficulties, as consumers have begun putting a higher importance on the value of each product they purchase which has in turn caused them to reassess the brands they regularly patronize. This trend bodes well for the future of travel for treatment post-COVID; with plentiful availability of cost-efficient care across the world, the competitive price tag of foreign medical and dental treatments will likely attract a new client base from this value-forward consumer culture.

Likewise, lockdown restrictions caused many services and consumer experiences that are required to be performed in a specific environment to be put on pause, causing global patrons to reevaluate their spending habits toward more home-bound goods and products. This has been reflected in international trends toward at-home wellness, well reflected in the United States' microcosm as health and fitness equipment revenue doubled in the country between March and October 2020 to $2.3 billion, with stationary bikes tripling in sales and treadmill equipment retailing at a 135% higher rate than before the pandemic (Shaban, 2021).

2.3 A TECHNOLOGICAL SHIFT TOWARD THE VIRTUAL WORLD

With global consumers largely limited from movement throughout the pandemic, technological evolutions looking to cater to this newly home-bound audience helped foster tech sector innovations that would normally take a number of years at an accelerated pace. Preestablished businesses, too, were forced to focus on their online presence to cater toward this rapidly growing online consumer base, particularly encouraged by adult consumers' upping their average mobile screen time by 60–80% across the pandemic.

> Overall digital device usage increased by 5 hours, giving a plunge to screen time up to 17.5 hours per day for heavy users and an average of 30 hours per week for non-heavy users. A recent study reported 8.8 h of screen time among younger adults and 5.2 hours among elderly (>65 years old). (Pandya and Lodha, 2021)

As such, the digitization of companies' customer interactions has been amplified across international regions with the average share of these interactions hitting 58% globally from a 36% prepandemic rate. The adoption acceleration of business-oriented digitized interactions was particularly potent in the Asian-Pacific area, where it moved forward by four years in advancements, while the European and North American regions each accelerated at a three-year rate (McKinsey & Company, 2021b). The overall digital offerings from companies increased at an even faster pace than digital customer interactions; the global market share of partially or fully digitized offerings by businesses went up to 55%, an increase from 35% prepandemic and an acceleration of a massive seven years faster than previously anticipated.

The workplace, too, experienced huge shifts toward digitization, as many workers that were fortunate enough to retain employment throughout their pandemic were forced away from the office and into working remotely. With forecasts estimating that 25–30% of the US workforce will retain this new work-from-home scenario into a postpandemic world, many technological evolutions that have allowed the transition to effective remote work output are expected to carry over in use into the future (Global Workplace Analytics, 2021).

According to a study carried out by the Institute of Global Health Innovation, this shift toward remote work brought on by technological innovations has actually increased worker efficiency far above prepandemic levels, making this virtual-forward world particularly enticing for businesses as well as consumers.

> Digital technologies and data solutions have increased the productivity of staff, ranging from 54% in the UK to 86% in the UAE.... A further 66% of global respondents agreed that their staff quickly adapted to using new digital tools since the onset of COVID-19. Around 61% of respondents agreed that their organizations are using these digital solutions more effectively than before the pandemic began and 58% reported that digital solutions have made operating models more efficient. (Thompson, n.d.)

This movement toward digital innovations has been particularly evident in the health industry, as pandemic-born needs for distanced care have resulted in the international telehealth use increasing by 38-fold and into a quarter-trillion-dollar sector. While this shift was experienced more strongly in areas like psychiatry and substance use disorder treatments, other speciality sectors normally confined to the boundaries of in-person businesses like cardiology, endocrinology, and neurology similarly experienced upticks on a smaller scale (Bestsennyy et al., 2022).

Now, with consumers experiencing the convenience of receiving quality care and medical advice from licensed professionals from the comforts of their own homes, compounded on top of the previously explored digitized innovations around the globe, this holistic movement toward ease of accessibility is anticipated to continue well into the future after COVID-19, establishing digital services as the new frontier of the modern global marketplace across all industries.

2.4 COMING TO CONCLUSIONS

The COVID-19 pandemic has undeniably changed the way the entire world works, bringing forth major shocks to nearly every modern industry. Between travel restrictions, job layoffs, technological advancements, and shifting consumer concerns, this external event largely derailed the prevailing trajectory of international economies and industries and reshaped the global marketplace as we know it, innovating new and preestablished products, services, technology, and travel to adapt to the new-normal and matching consumer demands.

This has been clearly demonstrated throughout the medical industry as a whole; while medical tourism and in-person care was largely limited by the pandemic, advances in telehealth brought forth by COVID-19 have created a more convenient and accessible healthcare market for the modern consumer. Now, people living in rural or remote areas can receive speciality care they may have otherwise been limited from due to distance from specialists—an issue largely fixed by the mass expansion toward telehealth. With 76% of consumers saying they have an interest in continuing using telehealth moving forward beyond the pandemic, consumer-driven innovations such as this coupled with other industry advancements, like huge leaps forward in vaccine technology, are poised to endure throughout the global medical marketplace for years to come (Bestsennyy et al., 2022).

The dental sector has likewise experienced widespread transformations throughout the time period, with telehealth similarly becoming a prominent player in the industry. New digitized platforms have allowed dental professionals to compare pricing for available materials necessary for relevant procedures and practices, making for a more transparent industry that's financial-equifying effects are likely to trickle down toward patients' price tags. However, as previously explored, economic impacts to the world heavily impacted the dental sector in turn, causing dramatic setbacks as patients were only going to the dentist for emergency care and all elective procedures were put on hold. Moving forward, professionals within this space will need to continue to adapt to new technologies and consumer demands moving forward to maintain competitive relevancy, particularly as the majority of dental practices are operated by one singular dentist (Attaie, 2020). Patients were too limited from receiving affordable care abroad due to widespread travel restrictions and social distancing measures, but as borders begin to largely open worldwide, the travel for treatment dental sector is expected to rebound in turn as consumers make up for time lost during the pandemic.

For the wellness industry, something the Global Wellness Institute defines as "the active pursuit of activities, choices, and lifestyles that lead to a state of holistic health," COVID-born economic hardships caused the industry as a whole to contract by 11%, dipping from a $4.9 trillion valuation in 2021 to a $4.4 trillion valuation in 2020. Given the service-forward aspect of many parts of the wellness industry, the sector as a whole naturally contracted as consumers were barred from in-person services and treatments by movement restrictions. But as the health and wellbeing of consumers has been brought to the forefront of societal psyche by the pandemic, and encouraging growth in certain internal sectors like the at-home fitness market, the global wellness economy is estimated to expand by 9.9% between 2020 and 2025, outpacing the anticipated international GDP growth of 2.6% throughout the same period. This in turn bodes well for travel for treatment and medical tourism as a whole, with this specific sector of the wellness industry expected to surge by a massive 20.9% from 2020 to 2025 as well (Global Wellness Institute, 2021).

KEYWORDS

- **medical tourism**
- **travel for treatment**
- **pandemic**
- **COVID-19**
- **healthcare**
- **consumer trends**
- **virtual healthcare**

REFERENCES

American Society of Plastic Surgery. *National Clearinghouse of Plastic Surgery Procedural Statistics. 2020 Plastic Surgery Statistics Report*, 2020. Retrieved from https://www.plasticsurgery.org/documents/News/Statistics/2020/plastic-surgery-statistics-full-report-2020.pdf

Arslan, U.; Cetin, K.; Tosin, Y.; Firat, Y. D.; Cayci, H. M.; Erdogdu, U. E.; Tardu, A.; Kurular, D. C. *Impact of the COVID-19 Pandemic on Operations in General Surgery Clinics in Turkey*; US National Library of Medicine, n.d. Retrieved on February 11, 2022, from https://pubmed.ncbi.nlm.nih.gov/34524121/

Attaie, F. *COVID-19: The Unintentional Accelerator for Dental Industry Transformation*; Dentistry IQ, December 2, 2020. Retrieved on February 13, 2022, from https://www.dentistryiq.com/dental-jobs/article/14188330/covid19-the-unintentional-accelerator-for-dental-industry-transformation

Bennett, J. *Fewer Jobs Have Been Lost in the EU Than in the U.S. During the COVID-19 Downturn*; Pew Research Center, April 15, 2021. Retrieved on February 8, 2022, from https://www.pewresearch.org/fact-tank/2021/04/15/fewer-jobs-have-been-lost-in-the-eu-than-in-the-u-s-during-the-COVID-19-downturn/#:~:text=Roughly%209.6%20million%20U.S.%20workers,despite%20having%20a%20larger%20population

Bestsennyy, O.; Gilbert, G.; Harris, A.; Rost, J. *Telehealth: A Quarter-Trillion-Dollar Post-COVID-19 Reality?* McKinsey & Company, July 22, 2021. Retrieved on February 11, 2022, from https://www.mckinsey.com/industries/healthcare-systems-and-services/our-insights/telehealth-a-quarter-trillion-dollar-post-COVID-19-reality#:~:text=COVID%2D19%20has%20caused%20a,to%20replace%20cancelled%20healthcare%20visits

Bjerke, L.; Rosenfeld, R.; Watts, M. *Impact of COVID-19 on Beauty & Wellness*; Robert W. Baird & Co., 2020. Retrieved from https://content.rwbaird.com/RWB/sectors/PDF/consumer/Impact-of-COVID-19-on-Beauty-Wellness.pdf

Buckley, P.; Barua, A.; Samaddar, M. *US Income Inequality after the Pandemic*; Deloitte Insights, November 10, 2021. Retrieved on February 9, 2022, from https://www2.deloitte.com/us/en/insights/economy/issues-by-the-numbers/covid-impact-on-income-inequality.html

Diaz, L.; Gonzalez, J. L. *U.S.-Mexico Border Reopens after 20 Months of COVID Shutdown*; Reuters, November 9, 2021. Retrieved on February 7, 2022, from https://www.reuters.com/world/americas/us-mexico-border-reopens-after-20-months-covid-disruption-2021-11-08/

Euromonitor International. *Consumers in 2021*, n.d. Retrieved on November 11, 2021, from https://www.euromonitor.com/consumers-in-2021/report

Gjerløw, H.; Jensehaugen, J. *Dealing with COVID-19 in the Middle East*; The Peace Research Institute Oslo, 2020. Retrieved from https://reliefweb.int/sites/reliefweb.int/files/resources/Gjerl%C3%B8w%2C%20Jensehaugen%20-%20Dealing%20with%20COVID-19%20in%20the%20Middle%20East%20-%20PRIO%20Paper%202020.pdf

Global Wellness Institute. *The Global Wellness Economy: Looking Beyond COVID*, 2021. Retrieved from https://globalwellnessinstitute.org/wp-content/uploads/2021/11/GWI-WE-Monitor-2021_final-digital.pdf

Global Workplace Analytics. *Work-at-Home after COVID-19*, October 16, 2021. Retrieved on February 11, 2022, from https://globalworkplaceanalytics.com/work-at-home-after-COVID-19-our-forecast

Guan, D.; Wang, D.; Hallegatte, S.; Davis, S. J.; Huo, J.; Li, S.; Bai, Y.; Lei, T.; Xue, Q.; Coffman, D. M.; Cheng, D.; Chen, P.; Liang, X.; Xu, B.; Lu, X.; Wang, S.; Hubacek, K.; Gong, P. Global Supply-Chain Effects of COVID-19 Control Measures. *Nat. Human Behav.* **2020,** *4* (6), 577–587. https://doi.org/10.1038/s41562-020-0896-8

Gwee, S. X.; Chua, P. E.; Wang, M. X.; Pang, J. Impact of Travel Ban Implementation on COVID-19 Spread in Singapore, Taiwan, Hong Kong and South Korea During the Early Phase of the Pandemic: A Comparative Study. *BMC Infect. Dis.* **2021,** *21* (1). https://doi.org/10.1186/s12879-021-06449-1

International Trade Administration. *Impact of COVID Pandemic on eCommerce*; International Trade Administration, n.d. Retrieved on February 10, 2022, from https://www.trade.gov/impact-covid-pandemic-ecommerce

JP Morgan. *How COVID-19 Has Transformed Consumer Spending Habits*, 2020. Retrieved on February 10, 2022, from https://www.jpmorgan.com/solutions/cib/research/covid-spending-habits

Kranz, A. M.; Chen, A.; Gahlon, G.; Stein, B. D. 2020 Trends in Dental Office Visits During the COVID-19 Pandemic. *J. Am. Dent. Assoc.* **2021,** *152* (7). https://doi.org/10.1016/j.adaj.2021.02.016

Kuehn, B. M. Africa Succeeded against COVID-19's First Wave, But the Second Wave Brings New Challenges. *JAMA Network* **2021,** *325* (4), 327. https://doi.org/10.1001/jama.2020.24288

Lim, B.; Kyoungseo Hong, E.; Mou, J.; Cheong, I. COVID-19 in Korea: Success Based on Past Failure. *Asian Econ. Papers* **2021,** *20* (2), 41–62. https://doi.org/10.1162/asep_a_00803

Luhby, T. *As Millions Fell into Poverty During the Pandemic, Billionaires' Wealth Soared*; CNN, 2021, December 7. Retrieved on February 10, 2022, from https://edition.cnn.com/2021/12/07/business/global-wealth-income-gap/index.html

McKinsey & Company. *Consumer Sentiment and Behavior Continue to Reflect the Uncertainty of the COVID-19 Crisis*; McKinsey & Company, November 19, 2020. Retrieved February 10, 2022, from https://www.mckinsey.com/business-functions/marketing-and-sales/our-insights/a-global-view-of-how-consumer-behavior-is-changing-amid-COVID-19

McKinsey & Company. *US Consumer Sentiment and Behaviors During the Coronavirus Crisis*; McKinsey & Company, December 14, 2021a. Retrieved on February 9, 2022, from https://www.mckinsey.com/business-functions/marketing-and-sales/our-insights/survey-us-consumer-sentiment-during-the-coronavirus-crisis

McKinsey & Company. *How COVID-19 Has Pushed Companies over the Technology Tipping Point—And Transformed Business Forever*. McKinsey & Company, February 18, 2021b. Retrieved on February 11, 2022, from https://www.mckinsey.com/business-functions/strategy-and-corporate-finance/our-insights/how-COVID-19-has-pushed-companies-over-the-technology-tipping-point-and-transformed-business-forever

Nasseh, K.; Vujicic, M. *Modeling the Impact of COVID-19 on U.S. Dental Spending*; American Dental Association, 2020. Retrieved from https://www.ada.org/-/media/project/ada-organization/ada/ada-org/files/resources/research/hpi/hpibrief_0620_1.
df?rev=80803bf948fb4011bd804b50602db8b6&hash=63FFC42DD30DB259EEDC74A9D1024B17

Pandya, A.; Lodha, P. Social Connectedness, Excessive Screen Time During COVID-19 and Mental Health: A Review of Current Evidence. *Front. Human Dyn.* **2021,** *3*. https://doi.org/10.3389/fhumd.2021.684137

Plümper, T.; Neumayer, E. Lockdown Policies and the Dynamics of the First Wave of the SARS-COV-2 Pandemic in Europe. *J. Eur. Public Policy* **2020,** 1–21. https://doi.org/10.1080/13501763.2020.1847170

Prasad, N. K.; Englum, B. R.; Turner, D. J.; Lake, R.; Siddiqui, T.; Mayorga-Carlin, M.; Sorkin, J. D.; Lal, B. K. A Nation-wide Review of Elective Surgery and COVID-Surge Capacity. *J. Surg. Res.* **2021,** *267*, 211–216. https://doi.org/10.1016/j.jss.2021.05.028

Shaban, H. The Pandemic's Home-Workout Revolution May Be Here to Stay. *The Washington Post*, January 8, 2021. Retrieved on February 11, 2022, from https://www.washingtonpost.com/road-to-recovery/2021/01/07/home-fitness-boom/

Simon, R. COVID-19's Toll on U.S. Business? 200,000 Extra Closures in Pandemic's First Year. *The Wall Street Journal*, April 16, 2021. Retrieved on February 8, 2022, from https://www.wsj.com/articles/COVID-19s-toll-on-u-s-business-200-000-extra-closures-in-pandemics-first-year-11618580619

Thompson, D. *Embracing Digital: Is COVID-19 the Catalyst for Lasting Change?* Institute of Global Health Innovation, n.d. Retrieved from https://assets.ey.com/content/dam/ey-sites/ey-com/en_gl/topics/government-and-public-sector/ey-embracing-digital-is-COVID-19-the-catalyst-for-lasting-change.pdf

US Bureau of Labor Statistics. *COVID-19 Ends Longest Employment Recovery and Expansion in CES History, Causing Unprecedented Job Losses in 2020: Monthly Labor Review*; US Bureau of Labor Statistics, n.d. Retrieved on February 8, 2022, from https://www.bls.gov/opub/mlr/2021/article/COVID-19-ends-longest-employment-expansion-in-ces-history.htm

United Nations Conference on Trade and Development. *Global e-Commerce Jumps to $26.7 Trillion, COVID-19 Boosts Online Sales*; UNCTAD, May 3, 2021. Retrieved on February 10, 2022, from https://unctad.org/news/global-e-commerce-jumps-267-trillion-COVID-19-boosts-online-sales#:~:text=The%20dramatic%20rise%20in%20e,report%20published%20on%203%20May

Vieira, P.; Mackrael, K. U.S.-Canada Border Restrictions: What to Know about Crossing Requirements. *The Wall Street Journal*, January 25, 2022. Retrieved on February 8, 2022, from https://www.wsj.com/articles/us-canada-border-open-11628539364

Wang, X.; Shi, L.; Zhang, Y.; Chen, H.; Jiao, J.; Yang, M.; Sun, G. A Comparative Retrospective Study of COVID-19 Responses in Four Representative Asian Countries. *Risk Manage. Healthcare Policy* **2022,** *15*, 13–25. https://doi.org/10.2147/rmhp.s334326

World Health Organization. *Shortage of Personal Protective Equipment Endangering Health Workers Worldwide*; World Health Organization, 2020, March 3. Retrieved on February 9, 2022, from https://www.who.int/news/item/03-03-2020-shortage-of-personal-protective-equipment-endangering-health-workers-worldwide

Zhang, Z. China's Travel Restrictions Due to COVID-19: An Explainer (updated). *China Briefing News*, 2021, December 29. Retrieved on February 12, 2022, from https://www.china-briefing.com/news/chinas-travel-restrictions-due-to-COVID-19-an-explainer/

CHAPTER 3

BALANCING THE PHYSICAL AND VIRTUAL CONNECTIONS

ABSTRACT

As the COVID-19 pandemic prompted consumers to explore new digital avenues for healthcare, the impact of interpersonal connection on the human condition has become increasingly evident. From elders left isolated without visits in their senior communities to workers forced out of their traditional office dynamics and into remote work, the COVID-19 pandemic has had major consequences on the mental and physical health of consumers of all ages. This chapter dives into the importance of human connection and mental health, some of the psychosomatic symptoms of the pandemic, the pros and cons of accessible virtual care, and the balance of both in-person and digital healthcare opportunities in the post-pandemic reality.

3.1 LONELINESS IN A HYPERSOCIAL WORLD

From the dawn of time, human connection has been a vital component for the development of civilization. While forming close relationships was typically reserved for the family unit in prehistoric times, ancient humans were forced to group together out of sheer survival necessity, with better chances of providing food and living to tell the tale if humans banded up for day-hunts instead of braving the wild at night alone. As inventions and advancements accelerated the growth of communities around the globe over time, being connected became not just a want, but a need, as a fully functional and civilized society became the benchmark of both progress and power that still exists to this day (Pennisi, 2011).

Medical, Dental, and Wellness Tourism: A Postpandemic Perspective. Mary Schreiber Swenson, PhD, and Amit Bansal (Authors).

Now, as we forge forward into the 21st century, human connection has become an altogether indispensable part of modern culture, with socialization and regular person-to-person interaction ingrained into the contemporary consumer marketplace and into our individual upbringings. This is clearly evident from birth, as environmental factors and actions taken by caregivers early on later influence a person's subsequent functionality in society upon reaching adulthood, with potential negative consequences in the workplace and education system specifically brought forth by inadequate lessons on the importance of human connection during youth (Cook, 2013).

However, as different cultures vary across the world, so has each region's stake in the importance of human connection; in a study comparing the sociological aspects Eastern and Western societies hold most important throughout the child-rearing process, the respective values placed on different areas of socialization shows distinct variation in global conceptions of human connectedness (Park et al., 2014). East Asian cultures predominately expected independence from their children, displaying the area's lessened importance on fostering social bonds, while Western cultures trended toward respect and unselfishness—two hallmarks of positive interpersonal relationships.

No matter the childhood conditioning, modern research has proven one thing time and time again: human connection is vital toward living a long and healthy life. With former US surgeon general Vivek Murthy famously stating that "loneliness and weak social connections are associated with a reduction in lifespan similar to that caused by smoking 15 cigarettes a day," isolation from support systems result in clear negative impacts to the physical condition and exposes vulnerabilities to conditions like anxiety, depression, antisocial behavior, and increased inflammation at the cellular level, while maintaining close relationships with others altogether correlates with a lengthened and happier lifespan (Seppala, 2017). With 33% of surveyed global adults classifying themselves as "lonely," the fraying social fabric of society has expanded into an international pandemic in its own right (Ortiz-Ospina and Roser, 2020).

As such, many modern humans have attempted to increase their connections with other people using recent technological marvels like the internet and social media to stay up to date with friends, family, acquaintances, and strangers' individual livelihoods alike. Unfortunately, the popularity of social media has resulted in one-sided parasocial relationships where people feel connected to a user's activities and pursuits without the latter's awareness of the spectator's existence, correlating directly with reports of loneliness (Jarzyna, 2020).

But as the grips of the COVID-19 pandemic forced people into their homes due to lockdown measures and social distancing mandates, social media became one of the lone beacons of hope for human connection during the difficult time period, increasing users' daily average screen time on a global scale in spite of social media's demonstrable faults and negative effects on the psyche. With social media users spending 61% more time on networking sites during the first wave of the virus than the year previous, the en masse shift to virtual connections seems to be reflective of humans' need and desire to continue interacting with one another regardless of presented physical limitations, also offering an increased opportunity to forge relationships with others located far across the map from one another (Fullerton, 2021).

Taking the pandemic-born increase in average screen time into account for business' operational practices, companies, national governments, and previously stagnant services began transitioning into the online world to cater toward this locked-down consumer. While we have previously explored major advancements in the world of telehealth, COVID-19 prompted the near-instantaneous move to digital across most aspects of life, allowing children to attend school online, important work meetings to be conducted remotely via video conferences, and various forms of therapy—both mental and physical—adapted to be performed from the comfort of patients' own homes (OECD, 2020).

This virtual growth was particularly potent in Southeast Asia, which added 40 million new regional users to the internet in 2020 alone. As one of the last frontiers of the emerging digital economy, the pandemic prompted 70% of the regions' population to become connected online, adding to online education's user base with a 55% share of new users, with the same trend reflected in grocery e-commerce, where 47% of consumers were new in the East Asian market (Hoppe, 2021).

The acceleration toward digital has perhaps found its most effective role in the continent of Africa; while only holding 22% penetration of the internet throughout the region prior to the pandemic, the shift online helped many African nations work toward leveling the playing field in the global virtual marketplace. According to data from the *The Mobile Economy Sub-Saharan Africa* report, approximately $130 billion of economic value was added to the sub-Saharan African countries in 2020 by mobile technologies and services, accounting for 8% of the region's gross domestic product for the year (GSMA, 2020). With technology powerhouse Meta investing considerable funding into 45,000 km subsea cable line 2Africa, which is set to connect regions in Europe, Africa, and Asia to the internet, and another $1 billion in cable infrastructure invested into the continent by Google, more than 474 million people in sub-Saharan Africa are anticipated to have access and regular usage of mobile internet by 2025, adding to the 33 million users in Africa that connected to the internet for the first time in 2020 (Mwakideu, 2021).

Despite a head-start on digital growth before the pandemic hit, developed nations have been no exception to the massive virtual acceleration prompted by COVID-19, particularly demonstrable at the government level. As part of South Korea's Digital New Deal, the country looks to invest more than $44.2 billion dollars to upgrade its economy and public infrastructure toward digital innovations and artificial intelligence, while Saudi Arabia has invested nearly $4 billion in its nationwide pursuit of the same advances. Countries like France, Spain, and Egypt have similarly allocated $8.4 billion, $158 billion, and $375 million respectively to stimulate technological growth, with other efforts toward modernization taken throughout state governments in both Canada and the United States, showing the virtual trend's extension onto an international scale (Sullivan, 2021).

3.2 MOVING FORWARD, MENTALLY AND DIGITALLY

With internationally increased opportunities to connect with one another on a digital level, and subsequently render traditionally in-person services online in the process, a number of intriguing questions have presented themselves. As we emerge from pandemic-caused lockdowns and social distancing, how much of this newly discovered and technology-run virtual world will endure in our society's future? Can the digital experience be an adequate platform for building and sustaining meaningful human connections? How much do physical presence and human touch factor into feelings of love and connectivity? While we may be far removed from definitive answers until the realities play out in real time, certain information can allow us to anticipate some of the potential ramifications across all sectors of the wellness economy as the global population works toward the balance of digital and physical connections going forward.

Consumers in the burgeoning post-COVID e-commerce marketplace seem to be trending toward virtual interaction, even with artificial-intelligence-powered response systems; according to HubSpot industry data, engagement with conversational bots has boosted by 75% from prepandemic levels, demonstrating the population's desire for virtual services with a conversational and connectivity aspect (HubSpot, 2022). Even events conventionally hosted in-person, like concerts and large business conferences, are making the digital transition, with the virtual event sector expanded to grow by 10-fold from $78 billion to $774 billion between the years 2020 and 2027 (Grand View Research, 2020).

In turn, the global telemedicine industry is expected to be boosted by these same consumer demands toward online accessibility, with a forecast to grow from 2020's $144.38 billion valuation to $636.38 billion by 2028. Considering locational limitations and increased ailments brought on by age, demands from the international rural population, and consistently growing senior community are expected to keep demand for telehealth high, particularly when coupled with the now-requisite convenience and value telemedicine has given to the general population (Fortune Business Insights, 2021).

Of course, the increase in screen time on trend with consumer habits normalized during the pandemic has resulting effects on users' health, both young and old, despite its enhancement of human connection throughout COVID-19's isolationary periods.

> While mindful (and regulated) use of digital devices is linked with well-being, excessive screen time is reported to be associated with a range of negative mental health outcomes such as psychological problems, low emotional stability, and greater risk for depression or anxiety. Negative consequences often result when digital use is impulsive, compulsive, unregulated or addictive.... sTechnology has had a profound impact on what it means to be social, challenging its overhauling essence. Furthermore, the coronavirus has atrophied the social skills of many individuals in the absence of peers. With most of us too used to interacting on digital platforms, we have missed the subtler nuances of what it means to exercise our social skills and etiquettes around people; which is only likely to surge especially when social distancing will fade. (Pandya and Lodha, 2021)

These issues have become particularly evident among the youth population, as children who were removed from access to their smartphones and other technology experienced difficulty in concentration and increased irritability, as well as increased reports of sleep problems and the development of near-sightedness. Considering our exploration earlier in the chapter on childhood development's impact on positive social connectedness later in life, conditioning children to become reliant on technology at a young age could potentially manifest into unforeseen consequences and issues in adulthood, which could—ironically enough—be treated with further screen time and the utilization of telemedicine.

Virtual schooling has presented an interesting convergence of the growing use of technology in a compulsory field; while social media usage can have a clear impact on mental health, online networking is not a necessity when compared to education. In the burgeoning digital education experience, teaching professionals have been presented with the COVID-born conundrum of how to create connections within their classroom and amongst their students when inherently limited by the online-nature of the experience. As such, teachers and professors have needed to increase their efforts toward connecting personally with their students to increase overall classroom efficiency rather than relying on isolating video lectures alone, showing that, even in the virtual school system, human connection remains a vital component toward quantifiable success (Glazier, 2020).

As efforts toward resuming postpandemic activities continue, the new digital normal is expected to be even more virtual-forward in 2025, as the population becomes increasingly dependent on the convenience and ease of COVID-necessitated technological innovations, particularly as they've permeated through nearly all essential sectors—education, work, health, commerce, and social interactions. Considering 47% of respondents to a survey conducted by Pew Research said they believe life will be *mostly worse* in 2025 than it was pre-pandemic, it's safe to say that a widespread feeling of pessimism surrounding technological reliance will endure hand-in-hand alongside the normalization of these digital platforms (Anderson et al., 2021).

While mental disorders already claimed a stake across 10.7% of the global population ahead of COVID-19, the pandemic's arrival only bolstered those numbers to new heights, as isolating effects and previously explored dependencies on social media resulted in surges in anxiety and major depressive disorders on an international scale, increasing by 53.2 million and 76.2 million diagnoses respectively during the pandemic's first year in 2020 and disproportionately affecting the global female population (Dattani et al., 2021).

> Among women, there were almost 52 million additional cases of anxiety disorders and more than 35 million additional MDD cases present in 2020,

> compared to 2019. Mental disorders were anticipated to have a greater impact on females as they are more likely impacted by the social and economic consequences of the pandemic, specifically as challenges like school closures and illness disproportionately required women—especially those who were more likely to be financially disadvantaged during the pandemic due to lower salaries and savings than their male counterparts—to fill the role of caretaker. This comes as no shock, as females generally are more susceptible to depression and anxiety disorders compared to males, with susceptibility starting before the age of 15. (COVID-19 Mental Disorders Collaborators, 2021)

Elders, likewise, have not been exempt from mental health harm throughout the pandemic, as seniors isolated in living communities catered toward their age range have been unable to see or spend time with loved ones due to health precautions, increasing seniors' susceptibility for depression, dementia, and overall cognitive decline and early death (The National Council on Aging, 2021). This is only contributing to the world's larger dementia issue, as 10 million new cases present themselves every year as the global population continues to age (World Health Organization, 2020). And with the self-reported happiness rates of countries across the globe showing an increasing happiness gap, as well as increased reports of worry, sadness, and anger rising by 27% from 2010 to 2018, internationally pervasive mental health and emotional challenges have become a distinct obstacle to the prosperity of the modern world (Newman, 2019).

While COVID-19 may have pushed technological advances to new levels of acceleration, so has the pandemic clearly accelerated the deterioration of the world population's collective well-being in tandem with what the US Center for Disease Control has coined as "mass trauma," exposing the wide gap between mental health and needs and available mental resources. This newly apparent vulnerability to mental distress and substance abuse, even without diagnosed mental illness, now offers a path forward to help meet these widespread needs and increase well-being for all with increased options and access to mental healthcare services.

Though a misplaced stereotype exists of psychotherapy and other mental health treatment options as only necessary for those with diagnosed disorders, research has proven the benefits of meeting with a professional to constructively talk through personal problems, both mundane and extreme, to find resolutions to issues and a path forward toward closure. Rather than encouraging emotional repression as is standard now, the mass adoption of therapy as an accepted treatment for all—regardless of diagnosis status—is expected to enhance the population's overall emotional awareness and expression in a healthier manner, with these effects hopefully compounding on happiness in turn (Ravenscraft, 2020).

This huge shift toward mental wellness and open discourse on the subject is just the beginning for a more conscientious world, opening up the doors for personal pathways toward higher levels of well-being, and similarly spurring on public health and community strategies to combat the growing mental health crisis, including 40% of US employers expanding their mental health benefits and similar policies reflected in the international workplace (Leonhardt, 2021). Between destigmatization efforts and open accessibility to professional care via telehealth, a more understanding perception of mental health afflictions has laid the stage for increased care attainability on the global stage moving forward beyond the pandemic. Still, aforementioned regional and cultural differences clearly persist throughout the mental health sector, as the United States' spending percentage composition on mental wellness was nearly five times the amount as that of China during 2020 at the pandemic's peak (Global Wellness Institute, 2022).

3.3 NEW SOLUTIONS

With mental health playing such a key role in the continued development of the global workforce, the youth community, and the overall population, the private sector has come back in full-force with answers to consumer demands by creating new solutions, services, and products to help people build resilience and improve their mental wellness. Full-on capitalizing on contemporary trends, businesses pivoting toward positively addressing mental health and offering adequate solutions to consumers have the dual-factor result of both stimulating the wellness economy and helping boost the population toward more accessible and comprehensive care.

Considering the rise of virtual schooling—a movement particularly strong amongst 90% of the world's (CITE) wealthiest nations—has kept children away from their classmates and increased feelings of loneliness, anxiety, and sadness among the school-age population, similarly causing a concerning jump in suspected suicide attempts among minors at the same time, particularly in the Asian continent (Vestal, 2021; Xinghui, 2021). In response, a number of virtual online counseling programs have arrived in turn with the aim of assisting and navigating children through their emotional struggles, processing the source of their feelings, and moving forward with a positive mindset (Reader, 2021). While there was a notable lack of pediatric health professionals prior to COVID-19, the overarching lessons learned from kid's pandemic-born problems are expected to see an increase in both existing children's mental health professionals and their subsequent availability for treatment through virtual channels, though this extended focus is expected to vary on a country-by-country basis and, as of now, seems to be concentrated in western countries.

Meanwhile, adults' standard workplace environment, likewise upended by the COVID-19 pandemic, has experienced a shift in tandem toward increased mental health accessibility. With an increasingly large number of the workforce—particularly Millennials and Gen Z-er's—placing a heightened value on mental health and taking power into their own hands to quit their mentally unfulfilling jobs, companies looking to retain employment have been tasked with establishing more adequate mental health resources for their staffers to benefit from. Between aforementioned efforts in the United States and on an international scale, like the 2021-founded Global Business Collaboration for Better Workplace Mental Health, the private sector has made a concentrated effort toward treating mental health issues in the workplace, looking to help remedy the estimated $1 trillion lost in the global economy every year due to lack of worker productivity while establishing a healthier, happier, and more efficient workforce worldwide in the process (Greenwood and Anas, 2021).

Mental health, too, is becoming a massive focal point of the overall built-in environment of the globe. On top of privately owned and public medical facilities (including dental practices) increasing the availability of their mental wellness services across the globe, infrastructure has adapted in turn to help assist the rising prominence of mental health issues in the population. With concerted efforts to construct buildings in a more user-friendly manner to help decrease stressors like overcrowding, noise pollution, and depressing design—all proven contributors to negative mental health—wisdom gained on the importance of the built-in home environment from the COVID-19 pandemic is anticipated to be reflected in the development of affordable and public housing internationally (Guha, 2020). Of course, countries like the United States, the United Kingdom, and other European nations with the viable resources to create or enhance these housing projects are likely to accelerate this built-in development toward mental health ahead of low-GDP countries, particularly those with preexisting overcrowding problems like India and Indonesia.

Alongside the private sector, a number of foreign governments have made a concentrated effort to ramp up their nations' telehealth services to provide their citizens' and residents' proper treatment for mental health issues and other health-related problems throughout and moving forward from the pandemic. Despite economic challenges, third-world countries like India, Sri Lanka, Brazil, and Colombia's governments' all established programs and policies toward the expansion of telehealth services, a trend replicated by first-world countries like New Zealand, the United Kingdom, the United States, Canada, and Australia (Basu et al., 2021). Most of this growth has been implemented around the World Health Organization's System Model Component framework, which has helped lead the international community toward this new era of efficient virtual medicine.

Though these advances toward telehealth undeniably assisted the global population throughout the pandemic, as the world returns to "normal," concerns remain among patients that these new, accessible forms of health services will fall by the wayside in favor of in-person treatment. This could provide issues for the elderly population or those with autoimmune diseases that are inherently mobility limited and could benefit from the continued use of telehealth, problems that arose throughout the United States when state health providers, such as in Maryland and Massachusetts, cut off these services after the end of their respective declared public health emergencies (Weiner, 2021). While private sector businesses can offer a viable alternative to public care, the driving factor of cost could alienate many low-income individuals in need of treatment from seeking care on these virtual platforms. As such, insurance providers expanding coverage to encompass telehealth permanently, continued intervention by national governments and reforms to medical licensing will be imperative for the progressive continuation of telehealth in the international marketplace.

Given hospitals clear prioritization of COVID-19 patients throughout the beginning of the pandemic, those with chronic conditions have similarly been forced to choose telehealth visits with medical professionals instead of in-office visits, as well as postpone elective procedures that could potentially alleviate or cure symptoms. While the long-term results of the pandemic in terms of an increase in chronic disease is unknown as of yet, chronic disease management has undeniably been disrupted by COVID-19 and its effects could potentially continue to affect this vulnerable population in the years to come even with modern telehealth in play (Hacker et al., 2021). However, regular access to telehealth has been a positive asset for those with diseases like diabetes who require constant check-ins without necessitating in-person visits.

The world's elders have similarly been affected by this switch to virtual care, especially when taking into account their lower propensity for properly harnessing new technology, though increased efforts to transition the senior population toward telemedicine has reportedly gone well. Considering a large percentage of elder's hospital visits and emergency care visits are avoidable with the input of a licensed medical professional, telemedicine has provided a cost- and time-saving alternative for elders needing an immediate opinion on their symptoms. According to a recent Canadian study conducted throughout the pandemic, "telehealth encounters were successful in resolving urgent and non-emergent needs in 84.0–86.7% of cases" among seniors, making the continued use of virtual services particularly germane for the globe's aging population (Bernstein et al., 2021).

Though many patients will surely take advantage of the return to in-person care, telehealth is expected to endure in the post-COVID marketplace, giving a more nuanced approach to contemporary medical care. According to the Global Wellness Institute, increased access and opportunities for health on the small scale, both virtually and in-person, will impact the wellness of the world population as a whole in turn—as "'me' is not safe if 'we' are not healthy."

> Wellness real estate starts with our individual micro-environments—the specific rooms where we sleep and work, our homes, and our workplaces or school buildings. But it also radiates more widely to our subdivisions, neighborhoods, communities, and cities. When we seek out health and well-being exclusively within our own micro-environments, we are living in a "wellness bubble." But the experience of COVID-19 shows us that our individual wellness bubbles cannot keep us safe and healthy. During the pandemic, some of us may have been sequestered in our homes; perhaps we ordered everything to be delivered, or perhaps we had domestic helpers to do our shopping and errands. But if delivery personnel, essential workers, and the people living in vulnerable communities are sick, then the risks increase for everyone. The pandemic has brought a new level of understanding that our individual health and well-being are intrinsically linked with our community and the broader environment. For example, a recent Harvard University study found that higher rates of air pollution are correlated with higher COVID-19 deaths in the United States. How we plan, build, and operate our homes and neighborhoods can have profound impacts on the social, economic, and environmental dimensions of community wellness, which in turn impact our individual wellness. (Global Wellness Institute, 2022)

As the world as a whole looks to move forward from the pandemic and back toward so-called normalcy, even some of the most mundane aspects of everyday life are anticipated to shift in line with lessons learned from the COVID-19 pandemic in hopes of building a better world in the process.

3.4 ASSESSING THE WORKPLACE'S "NEW NORMAL"

In a global survey conducted in March 2021, more than 70% of respondents were still working remotely out of office—a number just 2% down from September from the year prior (Statista, 2021). While many staffers have anticipated going back to working in-person, more than seven-tenths of the world's population still performing their professional duties from their homes has necessitated an overhaul in the way the international business community approaches online work—particularly as a lack of human connection has contributed to worker's previously explored mental health decline.

As such, telemedicine has become a key aspect of many employers' medical benefits to their workers, with most employers investing the money to expand their telehealth offerings beyond a niche feature and into an integral part of operations. This has been a net-positive for employees with chronic diseases, who have been able to get routine check-ins covered by their employers without having to leave the comfort of their own homes or their workday and has subsequently contributed to increased productivity (Palekar, 2021). These same fast, affordable and accessible components of telehealth have contributed to healthy staffers' acceptance of the virtual services as well, as convenience has become a hallmark consumer desire of the pandemic. And on the employers' end, if well utilized by staff, telehealth can end up causing considerable savings for companies, turning telemedicine's enhanced accessibility in the workplace into a win–win for all parties involved (Sammer, 2019).

This newfound importance on health in the workplace is expected beyond teleservices and into more comprehensive coverage packages for employees, which may see travel for treatment and medical tourism extend further into company culture farther than ever before. As employers continue to funnel more and more money into programs, services, and activities and equipment aimed at improving their employees' health and wellness, offering an à la carte treatment menu via travel for treatment can incentivize employees to save money for themselves and their employer by going to Centers of Excellence across the world to acquire the specialist care they need. By opening up the doors for medical tourism, employees are also exposed to a secondary opinion they may not otherwise be able to afford within their nations' borders, allowing for higher quality of care and better odds of tackling health problems from the correct perspective—an area that expanded telehealth coverage will also help improve.

In taking more time to research treatment options via telehealth, consumers across the world are given the ability to make smarter decisions based on their access to research information on medical and dental procedures. With employers assuming part of this knowledge burden by offering travel for treatment options to their staff, the international health market should become a more equitable place in turn, allowing the global workforce to acquire the quality care that they desire, at a price they can afford.

KEYWORDS

- **medical tourism**
- **travel for treatment**
- **pandemic**
- **COVID-19**
- **virtual healthcare**
- **mental health**
- **telehealth**
- **remote work**

REFERENCES

Anderson, J.; Rainie, L.; Vogles, E. A. *Experts Say the 'New Normal' in 2025 Will Be Far More Tech-driven, Presenting More Big Challenges*. Pew Research Center, February 18, 2021. https://www.pewresearch.org/internet/2021/02/18/experts-say-the-new-normal-in-2025-will-be-far-more-tech-driven-presenting-more-big-challenges/

Basu, A.; Kuziemsky, C.; De Araújo Novaes, M.; Kleber, A.; Sales, F.; Al-Shorbaji, N.; Flórez-Arango, J. F.; Gogia, S. B.; Ho, K.; Hunter, I.; Iyengar, S.; John, O.; John, S.; Kulatunga, G.; Rajput, V. K.; Ranatunga, P.; Udayasankaran, J. G. Telehealth and the COVID-19 Pandemic: International Perspectives and a Health Systems Framework for Telehealth Implementation to Support Critical Response. *Yearbook of Medical Informatics* **2021,** *30* (01), 126–133. https://doi.org/10.1055/s-0041-1726484

Bernstein, P.; Ko, K. J.; Israni, J.; Cronin, A. O.; Kurliand, M. M.; Shi, J. M.; Chung, J.; Borgo, L.; Reed, A.; Kosten, L.; Chang, A. M.; Sites, F. D.; Funahashi, T.; Agha, Z. Urgent and Non-emergent Telehealth Care for Seniors: Findings from a Multi-site Impact Study. *J. Telemed. Telecare* **2021,** 1357633X2110043. https://doi.org/10.1177/1357633x211004321

Cook, G. *Why We Are Wired to Connect*. Scientific American, October 22, 2013. https://www.scientificamerican.com/article/why-we-are-wired-to-connect/

Dattani, S.; Richie, H.; Roser, M. Mental Health. *Our World in Data*, August 2021. https://ourworldindata.org/mental-health

COVID-19 Mental Disorders Collaborators. Global Prevalence and Burden of Depressive and Anxiety Disorders in 204 Countries and Territories in 2020 Due to the COVID-19 Pandemic. *Lancet* **2021,** *398* (10312). https://doi.org/10.1016/S0140-6736(21)02143-7

Fortune Business Insights. *Global Telehealth Market Size [2021–2028] to Reach USD 636.38 Billion and Exhibit a CAGR of 32.1%*. Yahoo Finance, October 25, 2021. https://finance.yahoo.com/news/global-telehealth-market-size-2021-094800237.html

Fullerton, N. *Instagram vs. Reality: The Pandemic's Impact on Social Media and Mental Health*. University of Pennsylvania Health System, April 29, 2021. https://www.pennmedicine.org/news/news-blog/2021/april/instagram-vs-reality-the-pandemics-impact-on-social-media-and-mental-health

Glazier, R. A. *Making Human Connections in Online Teaching*. Cambridge Core, November 19, 2020. https://www.cambridge.org/core/journals/ps-political-science-and-politics/article/making-human-connections-in-online-teaching/B75E1072BF11501B25578D72B41E0167

Global Wellness Institute. *The Global Wellness Economy: Looking beyond COVID*; Global Wellness Institute, 2021. https://globalwellnessinstitute.org/wp-content/uploads/2021/11/GWI-WE-Monitor-2021_final-digital.pdf

Global Wellness Institute. *The Global Wellness Economy: Country Rankings—February 2022*. Global Wellness Institute, 2022. https://globalwellnessinstitute.org/wp-content/uploads/2022/02/GWI2022_GlobalWellnessEconomy_CountryRankings_Final.pdf

Grand View Research. *Virtual Events Market Size, Share & Trends Analysis Report by Event Type, by Service, by Establishment Size, by End-use, by Application, by Industry Vertical, by Use-case, by Region, and Segment Forecasts, 2021–2028* (GVR-4-68038-795-7), 2020. https://www.grandviewresearch.com/industry-analysis/virtual-events-market

Greenwood, K.; Anas, J. It's a New Era for Mental Health at Work. *Harvard Business Review*, October 4, 2021. https://hbr.org/2021/10/its-a-new-era-for-mental-health-at-work

GSMA. *The Mobile Economy Sub-Saharan Africa 2020*, September 29, 2020. https://www.gsma.com/mobileeconomy/sub-saharan-africa/

Guha, M. Mental Health in the Built Environment. *J. Ment. Health* **2020**. https://doi.org/10.1080/09638237.2020.1836559

Hacker, K. A.; Briss, P. A.; Richardson, L.; Wright, J.; Petersen, R. COVID-19 and Chronic Disease: The Impact Now and in the Future. *Prevent. Chronic Dis.* **2021,** *18*. https://doi.org/10.5888/pcd18.210086

Hoppe, F. *COVID-19 Helped Accelerate Southeast Asia's Digital Future. Now What?* Fortune, August 6, 2021. https://fortune.com/2021/08/06/southeast-asia-covid-digital-consumers-growth/

HubSpot. *Data: Industry Benchmarks from 100,000+ Companies*, 2022. https://www.hubspot.com/industry-data?category=total&topic=deal-pipelines&subTopic=deals-created&drilldown=null

Jarzyna, C. L. Parasocial Interaction, the COVID-19 Quarantine, and Digital Age Media. *Human Arenas* **2020,** *4* (3), 413–429. https://doi.org/10.1007/s42087-020-00156-0

Leonhardt, M. *Nearly 40% of Employers Expanded Mental Health Benefits*. Fortune, November 10, 2021. https://fortune.com/2021/11/10/mental-health-benefits-at-work-covid/

Mwakideu, C. *Can Africa Achieve Universal Internet Access by 2030?* Deutsche Welle, May 11, 2021. https://www.dw.com/en/can-africa-achieve-universal-internet-access-by-2030/a-59729090

The National Council on Aging. *COVID-Driven Isolation Can Be Dangerous for Older Adults*, March 31, 2021. https://www.ncoa.org/article/covid-driven-isolation-can-be-dangerous-for-older-adults

Newman, K. M. *World Happiness Report Finds that People Are Feeling Worse*. University of California Berkeley, March 20, 2019. https://greatergood.berkeley.edu/article/item/world_happiness_report_finds_that_people_are_feeling_worse

OECD. *Digital Transformation in the Age of COVID-19: Building Resilience and Bridging Divides*. Digital Economy Outlook 2020 Supplement, OECD: Paris, 2020. https://www.oecd.org/digital/digital-economy-outlook-covid.pdf

Ortiz-Ospina, E.; Roser, M. *Loneliness and Social Connections*. Our World in Data, February, 2020. https://ourworldindata.org/social-connections-and-loneliness

Palekar, S. *How Telehealth Enhances Employer-Provided Health Plans*. Benefits Pro, October 19, 2021. https://www.benefitspro.com/2021/10/19/how-telehealth-enhances-employer-provided-health-plans/?slreturn=20220128153328

Pandya, A.; Lodha, P. Social Connectedness, Excessive Screen Time during COVID-19 and Mental Health: A Review of Current Evidence. *Front. Human Dyn.* **2021,** *3*. https://doi.org/10.3389/fhumd.2021.684137

Park, H.; Coello, J. A.; Lau, A. S. *Child Socialization Goals in Western versus East Asian Nations from 1989 to 2010: Evidence for Social Change in Parenting*. Bryn Mawr College, 2014. https://repository.brynmawr.edu/cgi/viewcontent.cgi?article=1044&context=psych_pubs

Pennisi, E. *How Humans Became Social*. Science.org, November 9, 2011. https://www.science.org/content/article/how-humans-became-social

Ravenscraft, E. Why Talking About Our Problems Helps So Much (and How to Do It). *The New York Times*, April 3, 2020. https://www.nytimes.com/2020/04/03/smarter-living/talking-out-problems.html

Reader, R. *These Companies Want to Quell the Rising Kids' Mental Health Crisis*. Fast Company, June 16, 2021. https://www.fastcompany.com/90646891/childrens-mental-health-brightline-sprout-c8sciences-tech

Sammer, J. *Telemedicine Improves Health and Saves Money, If Employees Use It*. Society for Human Resource Management, August 16, 2019. https://www.shrm.org/resourcesandtools/hr-topics/benefits/pages/telemedicine-improves-health-saves-money-if-employees-use-it.aspx

Seppala, E. *Connectedness & Health: The Science of Social Connection*; Stanford Medicine: The Center for Compassion and Altruism Research and Education, June 27, 2017. https://ccare.stanford.edu/uncategorized/connectedness-health-the-science-of-social-connection-infographic/

Statista. *State of Telework COVID Worldwide 2021*. Statista, March 20, 2021. https://www.statista.com/statistics/1199110/remote-work-trends-covid-survey-september-december/

Sullivan, M. *Accelerated Digital Government*. Deloitte Insights, March 4, 2021. https://www2.deloitte.com/xe/en/insights/industry/public-sector/government-trends/2021/digital-government-transformation-trends-COVID-19.html

Vestal, C. *COVID Harmed Kids' Mental Health—And Schools Are Feeling It*. The Pew Charitable Trusts, November 8, 2021. https://www.pewtrusts.org/en/research-and-analysis/blogs/stateline/2021/11/08/covid-harmed-kids-mental-health-and-schools-are-feeling-it

Weiner, S. *What Happens to Telemedicine after COVID-19?* The Association of American Medical Colleges, October 21, 2021. https://www.aamc.org/news-insights/what-happens-telemedicine-after-COVID-19

World Health Organization. *Dementia*, September 21, 2020. https://www.who.int/news-room/fact-sheets/detail/dementia

Xinghui, K. How the Pandemic Brought Asian Children to Breaking Point. *South China Morning Post*, November 27, 2021. https://www.scmp.com/week-asia/health-environment/article/3157511/how-pandemic-brought-asian-children-breaking-point

CHAPTER 4

HOW THE INTERNATIONAL MEDICAL TOURISM SECTOR FOUGHT THROUGH THE PANDEMIC, AND THE INDUSTRY'S ROAD TO POST-PANDEMIC RECOVERY

ABSTRACT

With the COVID-19 pandemic severely limiting international travel, countries across the world were forced to adapt their medical tourism industries in turn. This chapter analyzes how the pandemic's peak affected major players in the sector and how they adapted accordingly, the reignition of the global medical tourism industry, and how widespread tourism restrictions shifted consumer travel trends toward global wellness experiences.

4.1 TRAVEL BEGINS TO REIGNITE INTERNATIONALLY

Now, two years after the inception of the COVID-19 pandemic, consumers across the globe are shifting in their pursuit of traditional care and healing to cater toward their wants and needs, particularly as health has become so paramount in the societal psyche throughout this time frame. This has resulted in many patients putting a focus on travel to find more personalized wellness solutions, taking advantage of reopened borders and reactivated global tourism sector to seek out healing on their own terms.

As previously explored, lockdown measures and travel restrictions brought onto the pandemic severely limited the movement of the global populus, forcing them to hunker down within their native borders, or wherever they

Medical, Dental, and Wellness Tourism: A Postpandemic Perspective. Mary Schreiber Swenson, PhD, and Amit Bansal (Authors).

resided at the moment, with little-to-zero flexibility for external travels. In turn, this has resulted in a population chomping at the bit to travel again, with 2021 surveys conducted by McKinsey & Company revealing this trend's presence on an international scale.

> A recent survey of 4,700 respondents from 11 countries around the world, conducted by the International Air Transport Association (IATA), revealed that 57 percent of them expected to be traveling within two months of the pandemic's containment, and 72 percent will do so as soon as they can meet friends and family. In our China travel survey, we see more and more respondents yearning for leisure trips further afield; 41 percent say they want their next trip to be outside China, the highest level we've seen, despite borders remaining sealed. (McKinsey & Company, 2021b)

According to the Travel Insights Recovery Dashboard, which tracks travel spending and data in the United States, US consumers spent $92 billion on travel in December 2021, an output on par with December 2019's prepandemic travel spending levels and showing consumers' demonstrated desire to get back out into the world. This same report similarly found that 8 out of 10 US citizens are excited to travel within the next 12 months as of the beginning of 2022, clearly showing the Western community's staunch commitment to making the most of eased-down restrictions.

Of course, while much of the focus of the newly reinstated travel trend has been put onto international excursions, domestic travel is similarly expected to experience a resurgence on the global landscape, particularly concentrated in countries with near-zero contamination rates like New Zealand and Australia. Though nations with this mindset may maintain border restrictions and bans on international travel (both coming in and going out) longer than their fellow first-world peers, consumers subsequently stuck within their native countries will be encouraged to explore their nation's domestic offerings in turn. This will also bring forth new stimulation in the domestic travel for treatment sector, as consumers eager to explore beyond their hometowns and seek out new wellness solutions can travel to find healing opportunities previously unknown to them in another nearby state or province.

Wellness travel on its own has become a decidedly important player in the international tourism sector postpandemic; a recent survey conducted by American Express revealed that 76% of respondents desired to spent more on wellness-focused travel, with another 55% saying they would be willing to pay additional costs for wellness-related activities on forthcoming vacations. With a reactivated global economy, a worldwide ease on travel restrictions, and an adventure-hungry consumer base ready to get back out into the world, the postpandemic marketplace has been primed for a surge in wellness-related travel on an international scale.

4.2 MEDICAL TOURISM AT THE PANDEMIC'S PEAK

Though the travel for treatment industry has had a major upward turn over the recent months, it's not to say that the demand for medical tourism was nonexistent during the pandemic. Quite the opposite, in fact, as widespread layoffs, sky-high employment rates and increased economic hardships made many consumers unable to afford the cost of adequate medical care.

> "We are seeing a pent-up demand for medical tourism during the pandemic, particularly in the United States where a fast-growing number of Americans are traveling across the land border with Mexico for health purposes," said David G. Vequist IV, the founder of the Center for Medical Tourism Research, a group based in San Antonio, Texas, and a professor at the University of the Incarnate Word in San Antonio.
>
> Even before the pandemic, millions of Americans traveled to other countries for savings of between 40 to 80 percent on medical treatments, according to the global medical tourism guide Patients Beyond Borders. Mexico and Costa Rica have become the most popular destinations for dental care, cosmetic surgery and prescription medicines while Thailand, India and South Korea draw in patients for more complex procedures including orthopedics, cardiovascular, cancer and fertility treatment. (Yeginsu, 2021)

Locations like Mexico, a perennially popular medical tourism destination for North Americans, still saw demand throughout the pandemic—albeit at a significantly lower rate—for the favorable quality-to-cost ratio of treatment available in the region, despite travel restrictions making it altogether more difficult to enter the country. While land restrictions, implemented between the United States and Mexico until November 2021, prevented many patients from traveling into the region by car, relatively open air travel—whose costs are offset by the low cost of treatment—has still allowed Americans and other international travelers to obtain medical and dental care in Mexico. With the land border reopened, Mexico's medical tourism sector is expected to experience a surge in patients in turn (Slotnik et al., 2021).

Turkey, too, has been able to establish a robust travel for treatment sector throughout the pandemic, shirking the travel restrictions implemented by neighboring countries to continually entice in foreign tourists after opening for business again at the beginning of July 2020 with its newly instated Safe Tourism Certification Program. Even throughout the Delta strain's prominence in the global caseload, Turkey refrained from requiring any COVID-19 tests on foreign inbound travelers, only requiring the tests for British citizens at the time. Similarly, Turkish nationals were given a strict evening curfew to follow throughout the Delta wave, while foreign tourists were allowed to roam the streets of Istanbul without restriction, making Turkey an even more attractive destination to cooped-up foreigners amidst the pandemic—especially for travel for treatment.

According to tourists in the area at the time, the pandemic provided a ripe opportunity for medical tourism thanks to a range of factors, including cheap hotels, low-cost airfare, altogether empty locations, and "down time in a foreign city where you don't have to explain to anyone why or how you got" surgery, recounted one British national who received a hair transplant in Istanbul at the height of the pandemic in December 2020. With the economic advantages of travel for treatment taken into account during this financially difficult time, it's no surprise that international patients, intent on bettering their health and wellness, continued to turn to medical tourism for care throughout the pandemic.

"It doesn't matter if you're in Europe or the Americas," medical tourism expert Paul McTaggart told the *New York Times* in January 2021. "Financially challenged individuals will seek out and be more receptive to the idea that you can travel abroad and see substantial savings for medical treatment and get past the line" (Yeginsu, 2021).

As a perennially popular tourism destination for both medical and recreational purposes, Croatia's warm weather, rich wealth of natural attractions, prime location on the Adriatic Sea, and balanced Mediterranean diet continued to attract international medical tourism patients throughout the pandemic. This was bolstered by Croatia's relatively low caseload when compared to other European countries, as the nation experienced 20 times less infections than its neighbors thanks to high levels of hygiene and cleanliness throughout the country. As such, Croatia became one of the first destinations to open its borders to the international community once more during the early stages of the pandemic, allowing near-free movement for travelers throughout the region and attracting medical tourists from across the world. Here, and in nearby Hungary, dental tourism began soaring over the pandemic's first summer as Europeans, plagued by dental appointment cancellations, flocked to the countries to receive dental care.

Other European countries were slower to allow in medical tourists, with Portugal waiting until August 2020 to extend access to its "Portugal Health Passport" to all international patients to boost the country's tourism recovery and instill a feeling of confidence to potential tourists. Looking to continue this feeling of goodwill, Portugal later announced its full-coverage COVID-19 travelers insurance, while other destinations, like the Republic of Cyprus, continued to keep its borders shut throughout the first wave despite insistence to reopen from other countries like Britain and Russia.

While select Western countries have clearly made a concerted effort to make themselves attractive to patients in the middle of a global pandemic, burgeoning medical tourism regions like Southeast Asia have similarly put an important stake in reinstating their travel for treatment industries. In Thailand, the Thai Ministry of Tourism announced in June 2020—just a few months into the pandemic—that it would give first-priority treatment to foreign patients who had pre-scheduled appointments in Thailand's private hospitals. While subsequent visa issues altogether limited the amount of foreigners who could enter the country, the Thai government's distinct decision to cater toward international patients shows the nation's clear commitment to strengthening its medical tourism sector, even throughout the peak of the world's largest international health crisis in a century. Likewise, the Thai government announced its intention of going after "quality" tourists over "quantity" in hopes that travelers with deeply lined pockets can help reignite the Thai economy.

To help promote Thailand's reignited medical tourism industry, the country launched the Medical and Wellness Resort of the World project in July 2020, bringing in more than 30,000 international tourists from China, Japan, and South Korea underneath a travel bubble to stay in the nation's top resorts in Phuket, Han Hin, and Krabi to engage in Thailand's available medical treatments and wellness attractions. The Thai government similarly announced an initiative to increase telehealth services for Thai people living abroad, increase the hospitality sector's collaboration with the medical industry, and launch B2B platforms extending the influence of Thai medical tourism on the international marketplace, with a goal of becoming a global travel for treatment hub by 2024. Later in 2020, Phuket announced the beginning of an \$89.5 million dollar medical plan for the Thai metropolis, constructing four healthcare facilities that will tackle rehabilitation, specialized elderly care and hospice programs, a hospital for chronic and terminal conditions, a cancer care center, and further funding allocated toward pandemic prevention. Thailand's concentrated campaign in promoting medical tourism had overwhelmingly positive results, with bookings of Thai hotels and wellness resorts experiencing 80–90% surges in bookings, both international and domestic, across a long weekend in the country at the end of July 2020.

Malaysia similarly seized its opportunity at the pandemic's peak to reopen its borders to medical tourists without limitations in an attempt to garner some much-needed economic stimulation for the country. Only needing to register with Malaysia's own Malaysia Healthcare Travel Council and take a pre-flight COVID-19 test, medical tourism patients to Malaysia were able to avoid quarantining upon arrival and were instead shuttled straight to their designated hospital for treatment or surgery, making Malaysia an intriguing travel for treatment destination mid-pandemic for those looking for streamlined care, particularly once it opened its doors to tourists from green-light countries like Australia, New Zealand, and South Korea over the summer of 2020.

In East Asia, technologically advanced countries like South Korea harnessed burgeoning technology to help bolster their travel for treatment sector against the pandemic's effects. Building and releasing a block-chain powered application with a focus on medical tourism, international patients can facilitate medical tourism-based transactions within the platform with a built-in international currency exchange backed by national banks. By digitizing inescapable parts of the medical tourism process, South Korea has positioned itself as a leader in modern travel for treatment innovations, allowing its foreign tourists to safely make transactions from the comfort of their own cellular devices and computers. Meanwhile, Taiwan began allowing international medical tourists into the country on August 1, 2020, allowing in travelers approved by the government via a newly implemented application for travel for treatment.

Despite some successes with medical tourism within the pandemic across the Asian continent, not every country in the region fared so well; India, for example, had more than 300 medical tourists stuck within the country at the onset of the pandemic, unable to return to their home countries due to canceled flights and newly imposed travel restrictions. While India had previously been projected to reach a $9 billion travel for treatment market by 2020, the pandemic put a major pause to that plan, with estimates that India experienced a huge $1 billion loss in medical tourism revenue during the first six months of the pandemic.

In the Middle East, Jordan began leading the way toward medical tourism recovery by allowing medical patients from fellow Arab countries to enter Jordan to receive treatment in June 2020, later launching its own website to help easily facilitate these travels for treatment opportunities. However, Jordan did keep two major safety factors in place for its visitors, requiring a negative preflight COVID test before entry and a mandatory two-week quarantine period once inside the country, served directly inside of the treating hospitals themselves. Regardless of these restrictions, Jordan was still one of the first Arab nations to reignite its travel for treatment sector while previously top-ranked medical tourism destinations like Saudi Arabia and the United Arab Emirates remained closed off during the time period, earning Jordan a stamp from the World Travel and Tourism Council as one of the few "safe travel" destinations across the world for medical tourism.

Still, the UAE still made efforts to prepare itself for postpandemic travel recovery, opening up a state-of-the-art $17.7 million plastic surgery facility—one of Dubai's top treatment sectors—in July 2020. This was soon followed by the announcement of Dubai reopening its borders to international travelers in the metropolis' continued efforts to become the Middle East's top medical tourism hub. Shortly after the announcement of Dubai's resumption of air travel and elective treatments, matched with the goal of bringing in 500,000 international medical tourists to the city in 2021, inquiries from international medical patients looking to receive treatment in Dubai shot up by 60%.

Standing in stark contrast, Iran's medical tourism revenue was cut by 95% by the pandemic throughout COVID-19's first wave, showing a variety of market reactions across the Middle Eastern region. While previously gaining major traction on the international marketplace for its robust medical tourism offerings, including $1.2 billion in revenue from travel for treatment between 2017 and 2018, Iran lost the majority of its upward trajectory at the beginning of the pandemic, with many of the region's hospitals receiving zero medical tourists during this time. Part of this can be explained by a lack of Afghan tourists, which compromised 48.8% of Iran's incoming medical tourism patients in 2019, due to enduring visa issues between the two countries.

By the end of the summer of 2020, medical tourism seemed to be back on the rise, with 22% of destinations having eased their travel restrictions compared to 3% in May of that same year. That being said, more than half of the world remained under lockdown, with 115 destinations—or 53% of all tourism destinations—maintaining closed borders throughout this time period (Jagyasi, 2020).

4.3 THE ROAD TO RECOVERY

Given the far-reaching effects of the COVID-19 pandemic, the road to medical tourism's recovery on a global scale has certainly not been linear, nor standard, across the international community. With unforeseen spikes in cases, highly contagious strains like Delta and Omicron wreaking havoc across the globe, and continued economic strains brought on by the virus, nations around the world have had to figure out their own unique paths to tourism recovery. Indeed, 20 of the countries that had opened their doors to international travelers in the first months of the pandemic were Small Island Developing States, which are generally dependent on economic activity driven by tourism. Still, another 41 of the 87 open destinations at the pandemic's peak were European, showing a demonstrable capability of reactivated tourism in Western countries, though most nations falling under this classification—including the United States—maintained a delay in reopening to nonessential travel, including medical tourism. However, once COVID-19 vaccines were introduced for public use in 2021's first-quarter, countries across the map began opening their borders for international travel of all types, bolstering travel for treatment back toward prepandemic levels.

But reinstating a tourism economy that had been unexpectedly grinded to a halt is no easy feat; take Thailand for example, one of the globe's top tourism destinations that was an early adopter of letting international travelers into the country through a preselection process, eventually fully opening its borders to all vaccinated travelers on November 1, 2021. Despite this head-start on the competition, Thailand still suffered massive blows to its tourism sector, with passengers on international flights dropping by 95% in September 2021 while only 9% of the country's hotels were filled at this same time—an issue only exacerbated by the emergence of novel coronavirus strains. When comparing total hotel guests and flight passengers in Thailand, year over year, Thailand actually had *more than double* the amount of tourists, both domestic and international, in 2020 at the pandemic's peak than it did throughout 2021, when countries theoretically had a better grip on the pandemic. This was certainly not a singular experience, as other tourism-dependent nations across the world similarly experienced dips in tourism in the pandemic's second year. Fortunately for Thailand, forecasts conducted by McKinsey & Company anticipate that, if the country can maintain a low caseload, Thailand could be back on track with its prepandemic tourism growth schedule as soon as 2024 (McKinsey & Company, 2021a).

Still, this anticipated resilience would not have been possible without Thailand's internal commitment to expanding its tourism sector, particularly domestically. As previously touched on, Thailand initiated the We Travel Together program in 2020 to encourage its population of 69.8 million people to travel within Thailand during the pandemic, providing travel 40% in subsidies to a total of six million nights of accommodation to build domestic travel's momentum, capped at $100 a night, as well as including other subsidies for food and beverage. Coupled with partial travel ticket refunds, the We Travel Together program became an undeniable success for Thailand, reaching its 6 million night goal, seven months after the program's launch, in February 2021. Considering the overt reliance of the Thai economy on tourism, this shift in focus toward domestic travel helped push Thailand through the pandemic—for a time, at least, as the Delta variant surge put a pause to these to We Travel Together in July 2021 for three months until October 2021.

Regardless of the We Travel Together program reaching its intended quota, domestic travel numbers were still less than half-that of 2019, dropping from $34.5 billion to $15.4 billion in 2020, resulting in the Thai government looking to reignite its foreign tourism sector for an injection of much-needed foreign money into the economy. As such, the country introduced a new program—the Phuket Sandbox—in hopes of prompting more international tourism to Thailand by offering travelers quarantine exemptions if they stayed in Phuket and creating a network of attractions available in the country, though the idea's positive economic impacts were only moderate as the Phuket Sandbox program unfortunately coincided with Thailand's experience of the Delta wave.

Though Thailand's own circumstances are unique, the country's overall path forward from the pandemic and tourism recovery can be applied on a larger scale, particularly as it has begun focusing on attracting "quality over quantity" tourists into the nation.

> Pre-COVID-19, China was one of the main contributors to Thailand's tourism income, accounting for more than 27% of 2019 tourism receipts. Given the current prudent approach of the Chinese government toward international travel, the road of return for Chinese visitors to Thailand will be a long one. China's international-flight seat capacity and passenger numbers remain down by 95% compared to pre-COVID-19 levels, and stringent public-health measures for international travel remain in place. Thailand, therefore, needs to reimagine its strategy and try to capture new

> sources of international travelers in markets where there are more rapid recoveries of international travel demand…
>
> …Recognizing these shifting traveler trends, and the resilient nature of premium traveler groups, the Thai government is striving to attract "quality" travelers from these source countries. Measures include revisiting and relaxing certain regulations—such as yachting regulations and taxes on personal belongings and luxury goods—to improve and stimulate the premium travel experience.
>
> Taking this a step further, the Thai government is preparing to launch a long-term residence program to attract foreigners to the country through new Long-Term Resident (LTR) visas (up to ten years), tax and investment incentives, foreigners' residential property ownership relaxations, and more. The program will target four key personas: the wealthy global citizen, the wealthy retiree, the work-from-Thailand professional, and the high-skilled professional. The country's ambition is to welcome over one million of these target personas and generate over 1 trillion baht in domestic spending in the next five years, beginning in 2022. (McKinsey & Company, 2021a)

Taking Thailand's approach as an example, nations across the globe looking to reignite their tourism sectors' revenue postpandemic have already begun following in Thailand's footsteps to stimulate their own tourism sector's recovery. Ideas like bundling hotel and flight packages together, implementing nuanced price strategies, and expanding activities and opportunities for wealthy travelers, much like Thailand, can help make countries more attractive options in the eyes of international tourists. Coupled with features like key partnerships—akin to Thailand's We Travel Together collaboration between the Thai government and private sector hospitality—to create a more comprehensive and trackable tourism ecosystem, an expansion in destinations and infrastructure, an enhanced digital marketing presence and financial support and subsidies provided by the government to the tourism sector, tourism-dependent nations can get back on track to their prepandemic tourism growth forecast.

As nations remain focused on reigniting their tourism sectors post-COVID-19—and, by extension, their medical tourism and travel for treatment industries—international medical tourism recovery is projected to reach new heights of prosperity within the next decade, even postpandemic. According to recent industry reports, the global medical tourism industry is anticipated to grow at an impressive compound annual growth rate (CAGR) of 16%

over the next several years to go from a $19.79 billion valuation in 2020 to reach $40.03 billion in 2025. Considering a CAGR between 5% and 10% is considered positive for large-scale industries, medical tourism's forecasted 16% CAGR is reflective of the travel for treatment sector's commitment toward recovery and provides an optimistic outlook on the future of medical tourism (PR Newswire, 2021).

4.4 AS MEDICAL TOURISM UPTICKS, PATIENT EXPECTATIONS SHIFT TO RECLAIMING THEMSELVES THROUGH WELLNESS, MEDICAL AND DENTAL PERSONALIZED EXPERIENCES

Now, as medical tourism begins to find its footing in a postpandemic world, consumers and patients across the globe are changing their expectations of healing and care toward more specialized experiences. Rather than pick one-size-fits-all experiences popularized over the years by all-inclusive resorts, consumers have begun trending toward destinations and attractions that specifically speak to their needs, creating a new era of medical tourism that will require countries to expand their variety of available wellness offerings in turn. Particularly after two years of minimal travel opportunities on a global scale, international travelers now have demonstrable intentions to embrace their travel opportunities to the fullest, with the intention to travel for leisure globally outpacing prepandemic levels in five major international countries according to figures from the Global Wellness Summit.

> Travel not only fulfills our sense of wonder, it allows us to connect with ourselves, our families and communities. "What's been reported about COVID-19 is the loss of the sense of taste and the sense of smell," said Cathy Chon, the founder of Hong Kong-based CatchOn, a FINN Partners company that specializes in travel marketing and communications. "What has not been reported about COVID-19 is the loss of our sense of time. Travel is one of the greatest timekeepers." (Global Wellness Summit, 2022)

Making up for lost time, travelers seem intent on instilling a sense of excitement and adventure into their wellness-related trips. As such, travel experiences tapping into an area's ancient wisdom are expected to rise in popularity, including in Australia's UNESCO World Heritage sites that have recently been handed over to the nation's indigenous Aboriginal population, allowing for a journey experienced through the eyes of Australia's natives. Select arctic boat tours are also expected to get an infusion of native culture, with Quark Expeditions new 199-passenger vessel featuring specially curated Inuit dinners made from locally foraged ingredients and native staples.

A rise in immersive experiences is similarly expected to occur in the new postpandemic wellness tourism market, particularly in the realm of agriculture and science as climate change's increased importance in consumer culture is set to be reflected in tourism trends. Now, hotels in Thailand, Scotland, Tennessee, and all across the world are allowing guests to forage for and prick their own produce through their meals, as well as learn more about concepts like irrigation, harvesting, planting, and sustainable farming. Similarly, travelers can expect increased opportunities to repopulate the destination's local wildlife, whether planting trees in the region or helping rebuild the ocean's coral reefs by bolstering coral nurseries.

Of course, given the massive stress to the population brought on by the COVID-19 pandemic, and lockdowns largely forcing people indoors, many consumers are keen to use travel as a key to reconnect—both with nature, and with themselves. As such, many hotels, resorts and travel companies have placed increased importance on offering opportunities for self-connection, grief and loss recovery, and positive mental health stimulation as part of tourism's new "need to serve … the whole person" (De León, 2021).

> "Hotels are really seeing people bring those new mind-sets, routines and preferences with them as they start to travel again," Ms. Klein said. "What that creates is a definite shift in expectations and experiences that hotels need to cater to, because they're not catering to the traveler from 2019…."
>
> ….Industry experts say the wellness travel trend is here to stay. The United States accounted for nearly 30 percent of the global wellness tourism market in 2020, and the sector is expected to grow to $919 billion by 2022 from $735.8 billion in 2020, according to the Global Wellness Institute. (De León, 2021)

With a reactivated global tourism industry back on its upward trajectory, the removal or lessening of travel restrictions across the globe, and a hospitality industry increasingly tuned-in to modern consumers post-COVID desires, medical tourism and travel for treatment are poised for a brand-new era of prosperity in the postpandemic world.

KEYWORDS

- **medical tourism**
- **travel for treatment**
- **pandemic**
- **COVID-19**
- **consumer trends**
- **wellness**
- **post-pandemic recovery**
- **international travel**
- **travel**

REFERENCES

De León, C. Travel as Healing. *The New York Times*, December 29, 2021. https://www.nytimes.com/2021/12/23/travel/wellness-travel-healing.html

GlobalWellnessSummit. *TheFutureofWellness2022*,2022. https://www.globalwellnesssummit.com/wp-content/uploads/2022/02/GWSTrendsFutureofWellness2022.pdf

Jagyasi, P. *Medical Tourism during COVID Crises*; Dr. Prem Medical Tourism, 2020. https://drprem.com/medical-tourism/medical-tourism-updates-in-covid-crisis

McKinsey & Company. *Reimagining Travel: Thailand Tourism after the COVID-19 Pandemic*. McKinsey & Company, November 30, 2021a. https://www.mckinsey.com/industries/travel-logistics-and-infrastructure/our-insights/reimagining-travel-thailand-tourism-after-the-COVID-19-pandemic

McKinsey & Company. *A Travel Boom Is Looming. But Is the Industry Ready?* McKinsey & Company, July 27, 2021b. https://www.mckinsey.com/industries/travel-logistics-and-infrastructure/our-insights/a-travel-boom-is-looming-but-is-the-industry-ready

PR Newswire. *Global $21.93 Billion Medical Tourism Markets, 2015–2020, 2020–2025F, 2030F—Cosmetic, Dental, Cardiovascular, Orthopedics, Bariatric, Fertility, Ophthalmic Treatments*, June 10, 2021. https://www.prnewswire.com/news-releases/global-21-93-billion-medical-tourism-markets-2015-2020--2020-2025f-2030f---cosmetic-dental-cardiovascular-orthopedics-bariatric-fertility-opthalmic-treatments-301310265.html

Slotnik, D. E.; Isai, V.; Abi-Habib, M. The U.S. Reaffirms Its Land Border Restrictions as Canada Relaxes Its Own. Mexico Has None. *The New York Times*—Breaking News, US News, World News and Videos, November 8, 2021. https://www.nytimes.com/2021/07/21/us/politics/us-border-canada-mexico.html

Yeginsu, C. Why Medical Tourism Is Drawing Patients, Even in a Pandemic. *The New York Times*, January 19, 2021. https://www.nytimes.com/2021/01/19/travel/medical-tourism-coronavirus-pandemic.html

CHAPTER 5

THE BENEFITS AND CHALLENGES OF TRAVEL FOR TREATMENT

ABSTRACT

While medical tourism can potentially bring patients high quality care at an accessible price point, traveling the world for medical care isn't without its challenges. From finding trustworthy facilities to properly vetting doctors ahead of time, this chapter explores the pros and cons of travel for treatment, provides patients with tips and tricks for maximizing preparation ahead of medical travel, and dives into travelers' necessities for a successful medical tourism trip.

5.1 THE PROS OF MEDICAL TOURISM

While medical tourism has been surging in popularity over recent years—a trend that's only expected to continue as the global industry is forecast to hit a $286.1 billion valuation by 2030 (GlobeNewswire News Room, 2021)—the conventional healthcare alternative has still been met with uncertainty for consumers widely unaware of travel for treatment's associated benefits, as well as risks. This can be explained in part by local healthcare providers and insurance companies looking to retain a hold on their client base, as the steadily expanding selection of medical tourism destinations and options has not been widely broadcast to the public. Still, with demand for travel for treatment continually on the rise, the pros of engaging in travel for treatment are becoming more and more evident to the international population. To simplify the understanding of medical tourism's comprehensive benefits, we will now explore how consumers from the United States—the world's

Medical, Dental, and Wellness Tourism: A Postpandemic Perspective. Mary Schreiber Swenson, PhD, and Amit Bansal (Authors).

largest spending healthcare market—can benefit from travel for treatment, pros that are similarly applicable to other developed countries around the world.

Medical tourism is becoming increasingly popular as people look for ways to save on medical costs, particularly after the economic challenges of the COVID-19 pandemic. The cost of diagnostic testing and medications is particularly expensive in the United States and in other first-world countries with nonsocialized healthcare, while the cost of pre- and postprocedure labor from health professionals is often dramatically lower overseas. Similarly, there is a high cost of malpractice insurance in the United States compared to the international community, while the quality-to-cost ratio of hospitals also tends to be much more favorable abroad. As a result, medical tourists can save anywhere from 25% to 90% on their procedure-related medical bills even with travel costs included, depending on the procedure they get and their destination country of choice.

Though insurance companies were once resistant to travel for treatment, providers are increasingly incorporating medical tourism into their programs. This is because both the insurer and insured can benefit from this type of symbiotic arrangement, as one of the party's savings directly results in further cost savings for the other, which has resulted in a number of providers offering medical bill discounts for patients pursuing travel for treatment. However, many providers won't provide care abroad unless in the case of an emergency, though the attainability of domestic medical tourism has undoubtedly been bolstered by insurance companies' new acceptance of travel for treatment.

Considering the major revenues generated by international tourists, medical tourism has become big business for many countries, inspiring them to bolster their offerings to travelers as a way to boost the local economy and health infrastructure. As such, patients traveling abroad for treatment often enjoy a luxury, spa-like medical tourism experience at a fraction of the cost they would pay locally for a sterile procedure process. International patients also can benefit from affordable one-on-one nursing care and an increased level of privacy, turning a medically purposed trip into a relaxing getaway in its own right, all while making major savings along the way.

In that same vein, many medical tourists seize the ripe opportunity to expand their medical tourism plans into a full-scale vacation, adding in extra time in the travel itinerary, before or after a scheduled procedure, to explore their travel for treatment destinations. This can make vacationing particularly inexpensive, as many insurance providers will cover the cost of flights to the procedure, or are included in packages put together by medical tourism agencies, especially when US patients elect to participate in the United States' domestic medical tourism sector. However, despite this undue opportunity to merge medical procedures with pleasure, patients will need to take the utmost care so as to not delay their recovery with the burden of additional tourism excursions.

Another strong benefit of medical tourism is access to the world's best and brightest surgeons, who oftentimes have a specialized surgical experience based on their country of origin, like Brazil with plastic surgery, Turkey with hair transplants, and Thailand with gender reassignment surgeries. Depending on the popularity of a procedure in a region, this trickle-down effect may also result in surgeries performed for near-free or low-cost in a given country. Plus, given the international reaches of the medical community, many of these specialized surgeons have received training and schooling on an international scale.

> It is often surprising to many medical tourists that their physician was trained in the United States. Not all physicians are, of course, but a surprisingly high percentage of them working in surgery abroad are trained in English-speaking medical schools and residency programs and then return to their home country. These physicians often speak multiple languages and may be board certified in their home country and a foreign country, such as the United States. (Whitlock, 2021)

Travel for treatment has also become an intriguing opportunity for the United States' growing immigrant community, which represents 19.1% of the total US population (Budiman, 2020), with many reasons as to why immigrants prefer to have treatments and procedures performed in their country of origin rather than within the United States itself. One reason is that, oftentimes, immigrants feel more comfortable communicating in their native language, eliminating any language barriers that could negatively impact the quality of care, while US immigrant patients pursuing medical tourism in their home country can be closer to friends and family for postoperative recovery purposes. Another all-important consideration for this population is cost—at least 25% of immigrants and noncitizen residents living in the United States are uninsured, compared to 9% of American citizens, a trend that unfortunately extends into the immigrant community's offspring.

Medical tourism also presents an opportunity for patients to bypass the recommendations of their local physician—though situations such as this should be approached with the utmost care and precision. Many doctors in local markets, like the United States, will refuse to perform certain procedures due to potential health complications, unsubstantiated research, or different national standards on what a good candidate for a specific treatment is. While doctors in foreign countries may be keen to perform procedures, a patient was refused in the United States, patients would be wise to listen to their long-term physician as a priority, as pursuing these treatments can result in risk or harm to the body.

5.2 THE RISKS OF MEDICAL TOURISM

On the flip side, traveling to a foreign country to receive medical treatment can also bring forth undue opportunity for things to go awry throughout the procedure process, especially when compared to locally performed procedures. While the major savings associated with medical tourism can encourage patients to rush into booking, it's important to both understand the risk associated with travel for treatment, as well as put the proper measures in place to protect traveler's against any unforeseen issues along the way.

For one, the quality of surgeons can vary greatly from country to country; this is because the level of training and experience requisite for doctors to become a specialist can differ significantly depending on their location. However, it is not always easy to determine who is a good surgeon and who is not, particularly given international language barriers; this is why Dr. Mary S. Swenson advocates for a marketplace that allows for transparent pricing, open borders, and options. To improve your chances of finding a quality surgeon abroad, take the opportunity to search online ahead of travel—this will give you access to information that you may otherwise not be privy to, and help ensure that you book treatment with a quality surgeon. It's also important to remember that even if the surgery is performed by a specialist in the area, there are no guarantees that everything will go smoothly, and that pursuing malpractice claims in a foreign country can be challenging, if not impossible, at times. Therefore, having ascertained your surgeon's credentials, qualities, and areas of specialty beforehand is an important way to remove risk from your travel for treatment journey.

> Another risk involved in Medical Tourism emanates from the fact that many foreign laws governing medical liability are not as strict as those in developed countries such as the US and Ireland. Compensation for medical malpractice can also differ and compensation for injury can be quite low in relative terms in other jurisdictions. Moreover, many foreign doctors

> may not have the same level of malpractice insurance. Patients may find themselves with very limited recourse, if any at all, for malpractice when it comes to the judicial systems in some countries. (Matheson, 2009)

On the other side of that same coin, the quality of nursing and technical staff can also vary in quality across the international community. While a well-trained nurse should be trained to identify and target problems in the pre and postoperative process, an inattentive or poorly trained nurse could be the make-or-break factor in a procedure's ultimate success. Taking additional time to research a hospital's staff alongside its doctors will help remove uncertainty about their nursing team's capabilities ahead of travel, while finding a former patient of the facility to similarly vouch for the staff is likely the best route for travelers to take when made available to them.

Likewise, it would be pertinent for travelers to check out a hospital facility's standards and rankings in comparison with its country's healthcare system. While some countries—like Turkey—have a robust healthcare system where public and private hospitals are purportedly up to the same standard of quality, patients would be wise to choose a private hospital over a public hospital in certain countries. While ensuring a hospital's international accreditation, as explored in the next section, can help prevent many issues, patients would be wise to search out photos and reviews of the facilities to see if it meets their standards, as well as confirm that the hospital has an intensive care unit in case anything does go wrong throughout the procedure.

Though pretravel quality assessments can do plenty to mitigate risk, a number of other unforeseen circumstances can occur throughout the travel for treatment process. For example, flying home soon after a surgery can expose patients to heightened risk of blood clots and infections, particularly on multihour or cross-ocean flights, which may prompt some medical tourists to extend their postoperative stay to lower this danger. Unexpected illnesses can also present issues, as traveling to new countries can expose patients to new viruses their body has no immunity to, or, as is common in 2022, to COVID-19, which would likely force travelers to enter an unexpected—and expensive—quarantine before returning home (Pacific Prime, 2021).

Fortunately, as medical tourism becomes an increasingly popular choice for international patients to receive medical care, insurance companies are beginning to offer more comprehensive coverage with medical tourism malpractice policies that can help protect travelers from unexpected risks, though, it is important to note, most providers will only issue these policies for patients booking doctors and hospitals accredited by international institutions (Center for Disease Control and Prevention, 2019).

5.3 PREBOOKING PREPARATION

With all the cost-saving potentials of travel for treatment taking into account, it is easy to see why people are traveling for medical procedures on an international scale. Considering that consumers can save a huge 90% in price on life-saving procedures like heart bypass surgery by looking outside of their country of residence for treatment, an increasing number of people are looking for healthcare alternatives through medical tourism. This trend extends far beyond invasive surgeries; an increasing number of patients are traveling for less serious treatments, such as dental work or cosmetic surgery. Yet no matter what level of invasiveness a treatment requires, medical tourists of all types need to understand the assumption of risk they are taking on when pursuing travel for treatment, as well as the pretravel steps they can take to maximize the flow of their medical tourism journey.

One simple way to bolster medical tourism's success is to look into the accreditation of a medical institution or facility prior to booking any procedures. By relying on a professional accreditation agency's third-party assessment of a facility, patients can help ascertain that a clinic will provide service up to the standard of the international community. Organizations like the Joint Commission International offer trustworthy insights into a hospital's capabilities, though a number of other reputable global accreditation services can offer a similarly accurate evaluation. Though many facilities are likely to display their international accreditation on their websites, available accreditation search engines can help simplify a patient's inquiry and confirm a clinic's reputation.

Centers of Excellence also offer an intriguing opportunity for medical tourists, considering their self-professed specialization in the treatment of certain medical issues and disorders. However, considering Centers of Excellence do not have a formal international accreditation process, it is vital for potential patients to thoroughly research a Center of Excellence's outside accreditations ahead of time rather than relying solely on the facility's word as bond.

On that same note, while a hospital's facilities may be subject to international accreditation, that same reputable, stamp-of-approval does not extend to a clinic's staff. Generally speaking, most accredited medical institutions will not subject their reputation to scrutiny by employing unlicensed health professionals. However, it is still important to research the individual doctors and surgeons that will be carrying out the surgery, whether through searching the internet or consulting with a doctor's previous patients for an in-depth review of their quality of care. Though there is no perfect science to researching a physician, a former patient's perspective can help provide extra insight into the potential success of a surgery as well as any unforeseen complications, helping raise any important red flags ahead of following through on the booking process.

Similarly, while a destination's tourism aspects—like beaches, a warm climate, or national landmarks—may intrigue patients to get out to see a location's sights after getting the green-light from their surgeon, medical tourists need to keep the sensitive nature of their recent procedure's recovery in mind when making postoperative plans. Exposing oneself to the sun by spending time by the pool or on the beach can result in the darkening of surgical scars and can turn a wound that would otherwise fade with proper care into a life-long part of the body. Likewise, preemptive water exposure could negatively affect stitches and sutures, so patients would be best advised to remain out of the water and avoid getting caught in the rain, and as such should thoroughly check destination's weather patterns and "rainy seasons" before booking to prevent scenarios such as this from occurring. And while booze may be requisite on a leisure trip, the substance can have adverse effects on postoperative medications and pain relievers, making alcohol a no-go during most patients' recoveries. For those looking to truly embrace the culture and nature of the city they are traveling to for treatment, these travelers would be wise to schedule in ample time ahead of surgery for tourist activities, or book their postoperative stay until their body has met a doctor's healing thresholds.

A simple approach to ensuring a comprehensive booking experience is to use a reputable medical tourism agency, such as leading global medical tourism agency MyMedChoices™. As the medical tourism industry has grown to new heights, so too has the number of agents who specialize in helping clients book surgeries. These agents are known as Medical Travel Facilitators and have a wealth of experience when it comes to arranging everything logistically, from hospital appointments to transfers and accommodation. Many of these agencies and agents have also inspected international facilities and met with the doctors in their network, meaning they can provide an unparalleled and accurate overview of each destination and its travel for treatment opportunities to clients and can help make sure patients' medical tourism journeys are set up to be both secure and streamlined throughout the booking process (Pacific Prime, 2021).

5.4 TRAVEL FOR TREATMENT ESSENTIALS CHECKLIST

While traveling for treatment has brought forth undue opportunities for cost-efficient medical care to the world population, the potential positive results of medical tourism could be completely lost if patients do not come to their unique journey fully prepared after the securing of requisite travel arrangements. Of course, as travel for treatment is a novel idea for many modern consumers, many medical tourists are unsure about exactly what they need to bring along with them on a medically focused trip that's different from standard leisure travel, prompting leading international medical tourism agencies like MyMedChoices™ to help prepare its clientele for their travel for treatment journey through specially curated packing checklists.

Though most medical tourism trips are focused on one singular patient's care, that is not to say it is a one-man journey; all medical tourists are suggested to bring a trusted travel companion along with them to help the travel for treatment process go smoothly and similarly provide an on-site contact in case of emergency. An added companion can also be particularly helpful with postoperative recovery, and therefore, as a pertinent addition to many patients' medical tourism plans, must be kept in mind throughout the travel preparation process.

To start with, medical tourism patients and their companions alike will need to ensure they have all their proper documentation in place to be received into their country of patronage. Though many nations hold an open border policy for certain passport holders—like US passport holders to Europe's Schengen area—not all countries have such lax entry requirements. Therefore, it is of the utmost importance for travelers to have properly acquired all requisite visas from destination countries' respective embassies prior to travel, ascertaining a streamlined entry and removing any unnecessary delays at customs along the way. Alongside visas, travelers should make sure to hold tight to their passports throughout their travel for treatment journey, as different countries have their own unique rules for tourists' identifications. Make sure to thoroughly review a destination nation's specific policies and limitations for international travelers prior to departure, and double-check that all legal documentation is up-to-date and not close to expiry.

Aside from travel documents, medical tourists should also prioritize bringing any and all relevant medical paperwork and records with them to their treatment destination. Considering how important a patient's medical history is to doctors crafting safe and effective treatment plans for their patients, thinking ahead and bringing these documents along could be a key factor in ensuring a surgery or procedure's ultimate success. Given the sensitive nature of these records, it's very important for travelers to make copies of any and all medical documents they bring on their journey in case anything goes wrong, with a recommendation to spread copies of the documentation across various pieces of luggage on the off-chance any suitcases go missing along the way. Patients should also make sure to have a list of emergency contacts close by at all times, keeping relevant information like the hospital, doctors, and hotels' addresses and contact information in an easily accessible location.

Keeping cash-on-hand in a destination's local currency is an undeniably important safety net to medical tourists, but while many travelers elect to perform their currency exchanges upon arrival, most travelers will find it more advantageous to handle currency exchange in their home market. By working with a trusted bank or institution in their area of residence ahead of time, travelers can make sure they get the best exchange rates possible rather than potentially face tourist-based upcharges upon arrival. Plus, having cash-on-hand straight off of the plan gives better flexibility for tourists to immediately get out of the airport and into the city, unburdened by post-arrival transactions. That being said, travelers can benefit most from this tip by acquiring currency in low denomination bills, as this will allow them to transact with local merchants, cab drivers and businesses with ease without having to break bills for change first.

For travelers' intent on using their credit or debit cards abroad, setting up travelers alerts with banks ahead of time can help eliminate any potential card blockages or transactional declines. Though many banks may still flag foreign transactions as suspicious for security reasons, letting financial institutions know *before* travel is vital for card-dependent medical tourists.

With the financial and documentation factors of traveling for treatment checked off the list, patients can turn their attention to after-care medications, as many countries do not offer the same variety of pharmaceuticals within their borders, and some seemingly standard over-the-counter medications may require a prescription or not be sold at all in foreign countries. Whether ibuprofen or aspirin, stocking up on choice medications and pain relievers ahead of time can help save patients a lot of trouble down the line in the recovery process, providing the comforts of home without relying on the potential cultural differences of another country's pharmaceutical industry. In this same vein, travel for treatment patients would be wise to bring some standard aftercare products along with them—like gauze, wound dressings, and band-aids—to create a robust postoperative kit for themselves.

To round off the ideal travel for treatment suitcase, prospective patients would be wise to include a variety of comfortable clothing for the journey, including loose clothing or clothing with buttons for ease of removal in a hospital setting. Patients planning for procedures above the neck should avoid packing any clothes that need to be put on over the head, as the fabric could snag on wounds or dressing and delay recovery or impede results. Slippers, pillows, and comforts of the home like bed linens are also a popular choice for many medical tourists, providing the comforts of home in postoperative recovery. Likewise, packing some nonperishable food items can help provide alternatives to local cuisine if it's not of a patient's taste, and including some key medications for food poisoning and upset stomach could aid with any uncomfortable adjustments to a country's available dishes (Solomon, 2021).

5.5 PREPARING FOR TRAVEL, BEYOND THE SUITCASE

Beyond the packing basics, MyMedChoices™ has presented an extended medical tourism checklist to ensure all travelers have dotted the i's and crossed the t's prior to departure. Here, MyMedChoices™ recommends medical tourists to check for any civil unrest or ongoing geopolitical events in their destination country of choice, as well as acquire a medical translator if they do not speak the local language. Decisions like these can help prevent tourists from unexpectedly getting caught in the middle of a country's internal issues, as well as make sure any and all questions they have about the process can be effectively communicated to their medical professionals without anything getting lost in translation along the way.

MyMedChoices™'s checklist continues to uplift patients' autonomy in curating their treatment by encouraging them to perform due diligence on their chosen doctors and hospital ahead of travel. Checking the credentials of a surgeon and facility and speaking with the chosen surgeon prior to departure are important factors to build trust for the patient, while making sure the hospital is aware of any food preferences and allergies beforehand and preprepping medical documents for transfer will simplify a patient's arrival, all while removing many potential kinks from the process.

That being said, nothing is quite as important as acquiring a second opinion from a local physician about the procedures or treatments planned to be performed on a medical tourism trip. Consulting with a local doctor, particularly one with knowledge of a patient's unique medical history, can help properly gauge all the relevant factors that go into travel for treatment and help potential medical tourists make an informed decision and is an imperative step in continuing with the medical tourism process. Plus, establishing a relationship between the international surgeon and a patient's local physician can help facilitate trust between all parties.

Beyond the requisite booking process of accommodation and transport, MyMedChoices™ goes on to encourage medical tourism patients to confirm pricing with the hospital before travel and secure the arrangements for postoperative nursing care, as well as fully inform themselves on the potential complications that could arise throughout the travel for treatment process. The travel for treatment organization likewise stresses the importance for all patients to acquire medical travel insurance on the off-chance that something goes wrong during a procedure or operation, making sure all patients have taken the steps to properly protect themselves.

An imperative aspect highlighted by MyMedChoices™'s complimentary checklist is the handling of postoperative medical records; while many patients may assume they will immediately have their medical records transferred back to them, nations across the world have different policies on the subject of medical records that will need to be explored by the patient ahead of time, particularly to check if the privacy standards between the country of procedures and patient's country of residence comply with one another.

5.6 A WORD OF ADVICE TOWARD POSTOPERATIVE CARE

Though many international health facilities—particularly Centers of Excellence—offer comprehensive postoperative care solutions, it is vital to ensure that a traveler's home country offers the same postoperative care opportunities to them upon return. While medical tourists should plan their itineraries to account for follow-up visits and check-ups with their surgeon in their place of treatment, certain aspects of postoperative care may endure past the end of the trip and is an important factor to consider ahead of embarking on a travel for treatment journey, particularly as poor postoperative care could negatively impact the positive results medical tourism is supposed to bring patients.

This likewise extends into specialized services, as some surgeons will recommend follow-up treatments that are not necessarily legal or green-lighted in native counties. For example, while many international health professionals will recommend treatments like stem-cell therapy for follow-up care, stem-cell therapy usage is banned in the United States outside of treatment for a select few rare medical issues (Pierson, 2021). While US patients can once again participate in travel for treatment and visit nearby countries like Mexico to receive treatment, where stem-cell therapy is approved and legal, this will undeniably add an extra cost to the treatment—one many medical tourists cannot afford, and a trend reflected in different variations across the international marketplace. Even for those patients who can accommodate the additional cost, adding unforeseen travel into the equation can be both exhausting and cumbersome for those who have already returned home from their initial medical tourism journey. Therefore, it's important for medical tourists to take into account the potential need to travel once more for postoperative care, or, as mentioned before, lock down a regional healthcare provider who can perform these specialized tasks related to recovery ahead of departure.

With the recent surge in telehealth's popularity during the pandemic, many medical tourists should now have the opportunity to converse and consult with their surgeon even after returning to their home countries through virtual platforms. However, as many patients feel more confident with consulting with an in-person physician about any lingering questions or follow-up consultations, medical tourists should make sure to find a local physician in their native country that can adequately assess their recovery progress to quell any potential problems down the line of recovery. That being said, many medical tourism destinations have put an emphasis on their postoperative care opportunities to help ascertain a patient's health before heading back home. This is an important feature for medical tourists from developing nations that do not have expansive medical facility infrastructure or routine access to top-quality specialists, who may elect for a longer postoperative stay to retain access to their surgeon throughout the recovery process.

Of course, as in the case of all medical procedures, treatments performed under medical tourism are not necessarily infallible. While there are millions of success stories from people who have indeed experienced positive results from medical tourism and are now staunch supporters of the travel treatment industry, there's no one-size-fits-all result for patients seeking cost-efficient care abroad, as participating in the medical tourism can produce imperfect results, and there is no guarantee that any individual traveler will not be one of the few dissatisfied with the outcome of their procedure.

Still, by following the pretrip preparation recommendations advised by the medical tourism industry's leading experts, medical tourists should have covered all relevant mitigating factors in advance, setting themselves up for the most success possible on their personal medical tourism journey.

KEYWORDS

- **medical tourism**
- **travel for treatment**
- **centers of excellence**
- **vetting**
- **quality of care**
- **international accreditation**

REFERENCES

Budiman, A. *Key Findings about U.S. Immigrants*. Pew Research Center, September 22, 2020. https://www.pewresearch.org/fact-tank/2020/08/20/key-findings-about-u-s-immigrants/

Center for Disease Control and Prevention. *Medical Tourism—Chapter 9—2020 yellow Book*, June 24, 2019. https://wwwnc.cdc.gov/travel/yellowbook/2020/travel-for-work-other-reasons/medical-tourism

GlobeNewswire News Room. *Medical Tourism Market Size to Hit US$ 286.1 Billion by 2030*. GlobeNewswire News Room, November 18, 2021. https://www.globenewswire.com/news-release/2021/11/18/2337536/0/en/Medical-Tourism-Market-Size-to-Hit-US-286-1-Billion-by-2030.html

Matheson. *The Pros and Cons of Medical Tourism*; GlobeNewswire News Room. Lexology, February 13, 2009. https://www.lexology.com/library/detail.aspx?g=66db0ce5-f789-4677-9699-07d0660cda79

Pacific Prime. *Medical Tourism: What You Need to Know before You Go*, March 22, 2021. https://www.pacificprime.com/blog/medical-tourism-what-you-need-to-know-before-you-go.html

Pierson, B. *Unproven Stem Cell Therapy Subject to FDA Regulation—11th Circ*. Reuters, June 2, 2021. https://www.reuters.com/legal/litigation/unproven-stem-cell-therapy-subject-fda-regulation-11th-circ-2021-06-02/

Solomon, B. *10 Essentials You Need to Pack for Your Medical Tourism Trip*. Medical Tourism Magazine, December 16, 2021. https://www.magazine.medicaltourism.com/article/10-essentials-need-pack-medical-tourism-trip

Whitlock, J. *Why Patients Are Turning to Medical Tourism*. Verywell Health, September 11, 2021. https://www.verywellhealth.com/understanding-medical-tourism-4069869

Travel for Treatment Planning Checklist

The key to a successful medical trip abroad, is to be prepared. Besides performing the cumbersome task of packing bags, it will be fruitful if you also invest some time to take care of various important medical travel essentials. We have prepared a checklist of all things you need to be sure about before leaving for your medical tour.

DESTINATION

- [ ] Have you selected the destination country?
- [ ] Have you done a background check on your destination country to check for any calamity, civil unrest, adverse political conditions, and any particular microbial infection that you might be more susceptible to?
- [ ] Are you familiar with the native language of your destination country? If no, then have you opted for a medical translator?

PROVIDER

- [ ] Have you selected the hospital?
- [ ] Are you satisfied with the credentials of the hospital and the surgeon?
- [ ] Have you spoken with your surgeon?
- [ ] Have you consulted with a local doctor or physician?
- [ ] Do you have your medical records prepared for transfer?
- [ ] Have you made your food requirements and preferences clear to the hospital?

TRAVEL & TRANSPORTATIONDER

- [ ] Have you received your visa?
- [ ] Have you booked your flights?
- [ ] Have you booked your hotel?
- [ ] Have you done tourism-related bookings?
- [ ] Have you arranged for the lodging of your travel companion(s)?
- [ ] Have you arranged for an airport pickup?

FINANCIAL, LEGAL & INSURANCE

- [ ] Have you confirmed the cost of the treatment with the hospital?
- [ ] Are you clear with the legal complications that might occur during the course of the treatment?
- [ ] Do you have your medical travel insurance?

POST-OP

- [ ] Have you opted for post-recovery nursing care? If yes, have you arranged for nursing care at your hotel or at the hospital itself?
- [ ] Are you clear with the post-treatment procedure to transfer your medical records back to you? Does it comply with the health privacy norms of both the countries?

CHAPTER 6

SELECTING THE RIGHT DESTINATION AND FACILITY

ABSTRACT

With dozens upon dozens of countries for medical tourists to travel to for care, selecting the right travel for treatment destination for a patient's specific needs requires the utmost care. This chapter provides a roadmap for patients on some of the top medical tourism destinations of today and their specialty procedures, as well as insights into conducting due diligence and the impact surgery can have on patients' mental health.

When it comes to the medical world, "perfection" in the procedure process can be hard to define. However, one factor that is essential for a perfect experience is feeling safe and confident in the care you will receive, including doing due diligence on the top-quality care available and having the right support system in place. With these preparations in place in advance, medical tourism can get as close to perfect as possible, particularly when it comes to choosing the ideal medical tourism for a patient's specific needs. After all, a meticulous selection of the destination is key to travel for treatment's ultimate success!

But with a seemingly ever-expanding array of countries to choose from, how is a medical tourist supposed to determine what exact destination and facility fits their unique needs best? Though a seemingly complex pursuit, diving deeper into the details of different facilities and locations—beyond previously explored standard preparation levels—is one of the most important parts of the medical tourism process.

Medical, Dental, and Wellness Tourism: A Postpandemic Perspective. Mary Schreiber Swenson, PhD, and Amit Bansal (Authors).

6.1 A PRIMER ON TODAY'S TOP DESTINATIONS

With so many medical tourism destinations available on the modern marketplace, a proper review of the specialties and offerings of various travel for treatment locations can make or break a patient's decision on where they choose to receive care. Without further ado, here's an alphabetical breakdown on the top medical tourism destinations to help provide important information on countries' medical tourism industries to potential patients.

Abu Dhabi: As the capital of the United Arab Emirates, this high-wealth metropolis has been outfitted with state-of-the-art facilities, particularly as the UAE looks to expand Abu Dhabi into one of the world's leading medical tourism hubs. Hosting a mix of private and public healthcare, most medical tourists would receive care in Abu Dhabi's high-end private hospitals. Though the quality of procedures in Abu Dhabi may be high, the country's high GDP has similarly increased the cost of treatment in turn and may not be the first destination choice for medical tourists looking to save money on medical care. Still, Abu Dhabi sits ninth on the Medical Tourism Index's global destination ranking for its consistently high-quality facilities and staff, though the destination is likely to appeal most to fellow residents of the Middle East (Medical Tourism Association, 2021). The city's warm climate and multi-time ranking as the World's Safest City may also be enticing factors to potential patients.

Argentina: A melting-pot of European and South American culture, Argentina has been steadily working its way up the ranks of the Medical Tourism Index to reach 20th overall ranking. Most procedures or services performed on medical tourists in Argentina are curative medicine, followed by cosmetic procedures and wellness treatments. While most of Argentina's medical tourism is concentrated in the capital of Buenos Aires, the industry's important contribution to the Argentinian economy has seen medical tourism expand throughout the country's cities for a variety of opportunities. Argentina is presently working on a revision of its tax structure to encourage more medical tourists to enter the country for healthcare through cost-savings benefits. While mainly catering to North American and Latin American medical tourists, Argentina's growing medical tourism infrastructure is expected to expand the country's reputation throughout the global travel for treatment industry (Thopson, 2022).

Armenia: Primely situated on the cusp of the European and Asian continents, Armenia presents an easily accessible location for medical patients from both of the highly populated regions to receive care, with a growing concentration in dental work. While still developing its core medical tourism industry, Armenia's rich cultural history and vast array of natural climates within its borders make Armenia an ideal destination for travelers hoping to fit in sight-seeing on their medical tourism journey, a factor only bolstered by Armenia's more than 200 therapeutic mineral springs that offer extra wellness opportunities to foreign travelers (Medical Tourism Magazine, 2021b).

Australia: As a native English-speaking destination, Australia has become a particularly attractive medical tourism destination for patients from North America and parts of Europe due to a lack of language barrier. With major landmarks and cultural sites to see—like the Sydney Opera House, the Great Barrier Reef, the Ayers Rock, and more than 30,000 miles of coastline—Australia offers travelers robust chances for tourism throughout their medical tourism journey. While medical tourism within the country has increased throughout the COVID-19 pandemic with expansions to Australia's domestic travel for treatment sector, Australians seem keener to participate in global medical tourism themselves than foreigners do to visit Australia itself for procedures. Regardless, medical tourism continues to grow throughout the nation and can offer a ripe opportunity for patients who care more about integrating tourism activities into their travel and removing language barriers than saving money (Leggat, 2015).

Austria: Located in the middle of the Alps, Austria has gained renown for its beautiful and picturesque European scenery, though it has yet to make a significant impact in the global medical tourism community and remains unranked on the Medical Tourism Index. Still, as Austria is equipped with numerous natural hot springs, holistic wellness clinics, and centuries-old architecture, the country offers travelers access to high-end wellness-related travel and extensive tourism. Likewise, Austria's high standards for the training of Austrian healthcare professionals have made specialty procedures relating to oncology and orthopedics one of Austria's main medical tourism attractions, though the high cost of care in Austria will off put many travelers from seeking treatment there despite Austria's visaless entry for citizens hailing from the United States, Korea, Australia, New Zealand, Israel, Japan, Singapore, Malaysia, and the rest of Europe (Medical Tourism Marketplace, n.d.-a).

Bahamas: With pristine beaches and clear blue water, travelers looking for some pre and postoperative rest and relaxation will likely gravitate to destinations like the Bahamas—particularly travelers from close-proximity destinations like the United States. That being said, the Bahamas has failed to attract international medical tourists as a result of failed programs and barriers enacted by the local government, and even has trouble retaining its own citizens for healthcare purposes, as more than 40% of Bahamians flock to nearby Florida for treatment instead of their own native country (IMTJ, 2021).

Bahrain: Gaining more and more exposure on the international circuit from its annual Formula 1 Grand Prix, Bahrain has solidified itself as the 36th top medical tourism destination in the world. Sandwiched between Qatar and Saudi Arabia, the island archipelago nation has allocated a high budget to expanding its tourism sector, including a 100,000 m^2 water park, meant to attract further foreign travelers. As such, growing its medical tourism industry has been one of the nation's development concentrations and has blossomed into the Persian Gulf's leading travel for treatment hub. However, a low physician-to-citizen ratio and a low healthcare per capita spending output may make travelers outside of the Middle East think twice about choosing Bahrain as their go-to destination (Medical Tourism Magazine, 2021a).

Barbados: Rich in sunshine, scenery, and both African and British cultural influences, Barbados has yet to implement an active medical tourism industry within its borders. Considering the global trend toward medical tourism expansion, Barbados is likewise expected to grow its travel for treatment sector in the years to come as the island nation's natural wonders make it an ideal destination for wellness-related tourism.

Brazil: The Medical Tourism Index's 28th top destination, Brazil has gained worldwide renown for its prevalence of high-quality plastic surgery and cosmetic procedures and high-number of Joint Commission International accredited facilities, though prices may vary depending on a surgeon's prominence. Medical tourists intent on sight-seeing will also benefit from traveling to Brazil for treatment with an extended itinerary, as the South American country offers warm weather, ample tourism activities and a variety of destinations, both rural and urban, to visit as the fifth largest country in the world.

Canada: Thanks to its top level medical professionals and facilities, Canada has secured the Medical Tourism Index's number one spot as the world's leading medical tourism destination. With healthcare costs nearly half that of the nearby United States, including for medications and prescription drugs, and highly trained professionals at a more favorable medical staff-to-population ratio than its southern neighbor, Canada is expected to remain at the top of medical tourism rankings for years to come. Not to mention the country's sprawling territory and wide variety of natural and urban destinations to visit, which can add an extra element of travel opportunity to Canada's already leading medical tourism industry (Find Private Clinics Canada, 2022).

China: Though China's healthcare system is predominantly public, a growing private healthcare sector is increasing China's role as a player in the international medical tourism market, currently sitting at the MTI's 33rd place. With low wait times, high-quality physicians, affordable procedures, and advanced technology, China's international reach via medical tourism is already expanding, with China's most common medical tourism procedures including cancer treatment, Parkinson's treatment, stem cell therapy, cataract surgery, and obesity surgery. That being said, China's huge language barrier, lengthy travel times, and consistently busy hospitals may detract from medical tourists choosing China as their travel for treatment destination.

Colombia: Much like Brazil, Colombia has built a reputation for its active plastic surgery industry, a factor only bolstered by its ranking at 22 on the World Health Organization's top 100 health systems—above that of both the United States and Canada. Featuring super specialty hospitals throughout the nation with services offered at an affordable price point, Colombia has become a leading destination for medical tourists looking for a cost-efficient solution to complex treatments, including neurology and cardiovascular procedures. Plus, a strong tourism sector will encourage medical tourists who look to supplement their journey with sight-seeing to visit the country, though Colombia's inadequate malpractice laws may give some travelers pause (Medical Tourism Marketplace, n.d.-b).

Costa Rica: Nestled at the intersection of North and South America, Costa Rica has become one of the most popular medical tourism destinations of the modern day, earning the MTI's sixth spot. A short flight from the United States, Costa Rica's lush territory and beautiful beaches have made the Central American ideal for both relaxation purposes and an abundance of tourist activities. Coupled with low costs, Costa Rica has become one of the world's leading destinations for dental work, cosmetic surgeries, noninvasive procedures like botox, and substance rehabilitation (Medical Tourism Marketplace, n.d.-c).

The Czech Republic: With affordable pricing, a history of high-quality healthcare, and a convenient location to European medical tourists, the Czech Republic has been able to offer cost-efficient treatments backed by cutting-edge technology to international medical tourists that have brought the nation to the 15th ranking on the Medical Tourism Index. While formerly high crime rates previously prevented some travelers from entering the country, crime has fortunately been on the decline in the Czech Republic over the past few years, attracting even more tourists to the nation in turn (Royal Medical, 2021).

The Dominican Republic: While the Dominican Republic has ranked highly on the MTI—sitting 19th on last ranking—a number of fatalities relating to medical tourism procedures performed in the country has set the island nation back in terms of its global reputation (Medical Travel Quality Alliance, 2015). That being said, with the Dominican government announcing its intentions to further develop the quality of the country's medical tourism industry, international tourists can expect the nation's active dental and esthetic medical tourism market to maintain its popularity moving forward (IMTJ, 2022).

Dubai: Bolstered by recent high-level investment from the UAE's ultra-wealthy population, Dubai has surged into the eighth leading medical tourism destination according to the Medical Tourism Index. Already hosting a strong tourism economy, Dubai's private healthcare sector has begun catering toward international medical tourists in spades, including launching a state-of-the-art online platform where travelers can find and book procedures from the comfort of their countries of residence. Now, orthopedic, dermatology, and ophthalmology procedures, as well as dental treatments, fertility treatments, and cosmetic surgeries lead Dubai's most sought-after treatments for medical tourists (Gulf News, 2018).

Egypt: With a 4000-year history and the tourist sites to match, Egypt has long been a top international tourism destination, but now looks to expand this sector into becoming one of the world's leading medical tourism destinations. In a recently announced initiative in 2021, Egypt has committed to expanding its medical tourism industry to further welcome international travelers, with the goal of expanding its influence in the international travel for treatment marketplace above its present 26th ranking on the MTI (Abu Zaid, 2021). With a number of natural treatment options—black sand therapy, mineral springs, and sulfuric springs—that help ease aches and pains, as well as 11 JCI-accredited hospitals and a superb cost-to-quality ratio, Egypt's comprehensive medical tourism programs offer a little bit of everything for the medical tourist looking to do it all (Vaidam, 2022).

France: Though a still burgeoning sector of the French economy, France's medical tourism offerings have been enough to rank the European country at 11 on the Medical Tourism Index. With easy access from Europe, North Africa, and the Middle East, the French's top-ranking public healthcare system has attracted a wealth of patients from across the world, particularly as many Europeans can receive care at low cost or for free in France by using the European Health Insurance Card. With quick turnaround service from private clinics, transparent pricing, and high-quality surgeons trained to rigorous French standards, it makes sense that France has become a major travel for treatment destination for cosmetic procedures, bariatric surgery, and cancer treatments, though the country's high price tag on medical procedures and complex paperwork system may detract from the country's benefits as a destination for some medical tourists.

Germany: The so-called heart of Europe, Germany has been a medical destination for Europeans for centuries, a role it continues to assume as the MTI's eighth top medical tourism destination. Primarily serving patients from Eastern Europe, Germany has become known for its favorable physician-to-population ratio and high-tech facilities, as well as its high standard of excellence for surgeons and medical staff, all of which have made Germany prominent for special procedures in neurosurgery, cardiac surgery, orthopedic surgery, and oncology. However, Germany's high cost of treatment and minimal focus on privacy may not lend themselves to travelers from the United States or England who are used to privacy throughout the healthcare process, as some German physicians will even decline to offer their patients hospital gowns or towels to cover the body during examinations and treatments ("Medical Tourism in Germany").

Greece: While Greece has regularly ranked at the top of WHO global healthcare system rankings for its blend of private and public healthcare, medical tourists would be wise to seek out care in one of the country's private clinics, which have both a lower wait time and better quality results. Between dental care, cosmetic surgeries, cancer, and fertility treatments, the Greeks' quality facilities, proficiency in English, international accreditation, and beautiful Mediterranean backdrop (a golden ticket for tourism) have made Greece into the Medical Tourism Index's 34th top medical tourism destination. On the other side of the same coin, issues like Greece's ongoing economic crisis and the country's inherent level of disorganization may prove to be detrimental for medical tourism activity, though the culture and beauty of the Greek Isles, plus an active party scene, could offset these cons (Medical Tourism Marketplace, n.d.-d).

Hungary: Modeled after the British healthcare system, Hungary has been able to offer its citizens high-quality care without the Brits' requisite wait times, a practice it extends into its up-and-coming medical tourism industry. Thanks to a concentration of patients from high-wealth countries like Germany and Austria, Hungary has outfitted its clinics with the state-of-the-art technology expected by these nearby medical tourists, making Hungary an even more attractive destination for travelers on an international scale. Well known for its popular dental sector, Hungary is also gaining prominence for plastic surgery and esthetic procedures and is well equipped to perform the large majority of standard medical procedures as well as certain specialized surgeries. With low-cost procedures and close proximity to other European countries, Hungary's popularity has positioned it at 23rd on the MTI rankings, though a lack of widespread accreditation and minimal malpractice protections may make the country a less attractive medical tourism destination to travelers located outside of Europe or the Middle East (Medical Tourism Marketplace, n.d.-e).

India: With procedures between 40% and 80% lower than its Western peers, India has grown into the MTI's 10th top medical tourism destination over recent years, offering hospitals at the same standard quality of American and British healthcare facilities that are likewise internationally accredited. Making things easier on medical tourists, the large majority of Indian medical professionals speak English, altogether eliminating the language barrier, bolstered by minimal waiting times for procedures. That being said, medical tourists should steer clear of small-scale clinics that likely do not have the same accreditation as the country's major hospitals, and be aware of lack of infrastructure, minimal malpractice protection, and differing cultural expectations in India that could all factor into a patient's destination decision (Medical Tourism Marketplace, n.d.-f).

Israel: Hosting one of the most advanced healthcare systems in the world backed by cutting-edge technology and a high specialist-to-population ratio, Israel has cemented its reputation as a modern medical tourism leader thanks to popular procedures in oncology, orthopedics, neurology, cardiology, and numerous other complex surgeries and treatments. Many patients will be attracted to Israel's state-of-the-art facilities, minimal language barrier due to a prevalence of spoken English in the country, and warm climate that have positioned it at eighth on the MTI, though higher pricing than nearby Asian countries and the country's ongoing geopolitical conflicts could detract from Israel's appeal for some patients.

Italy: With an abundance of culture, history, and beautiful scenery, Italy has been one of the world's most popular tourism destinations for years—and with the 18th place on the Medical Tourism Index, is beginning to make a major name for itself in the world of medical tourism. JCI-accredited hospitals, a minimal language barrier, and high-quality surgeons have all pushed Italy to hold a 17% stake in Europe's medical tourism industry, though Italian pricing is less competitive than other European medical tourism destinations (HTI Conference, 2017).

Japan: The globe's third top destination according to the MTI, Japan has become known for its specializations in cancer treatment, regenerative medicine, and heavy particle radiotherapy. Though more concentrated on domestic travel for treatment, Japan's state-of-the-art hospitals and highly trained staff, plus government investment, aim to bring Japan toward becoming Asia's top travel for treatment hub (Medical Tourism Magazine, 2021c). Plus, Japan's vast array of tourism attractions and variety of climates could inspire patients to add extra travel days to their medical tourism journey, though Japan's healthcare costs remain higher than that of other Asian countries.

Jordan: The Middle East's top healthcare system according to the World Bank, Jordan's mixed public-and-private healthcare system has attracted patients from across the globe, but travelers would be wise to seek out a private institution if choosing Jordan as their destination of choice. With nine JCI-accredited hospitals, medical malpractice policies, and a low cost of care, Jordan has reached 36th place on the MTI, particularly as it offers some of the lowest cost organ transplants in the world (Medical Tourism Marketplace, n.d.-g).

Korea: Well known for its bustling tourism sector, Korea has also cemented itself as a leading medical tourism destination over recent years, earning the Medical Tourism Index's 14th spot. Once known as the world's leading spinal care destination, Korea is growing its reputation for cosmetic procedures, dental care, organ transplants and cancer treatment, all backed by Korea's incredibly high standards of medical certification, technological advancements, and the JCI accreditation of 23 of Korea's private hospitals. With both tourism opportunities and quality of care offered in abundance at an extremely low cost, the language barrier is one of Korea's few detractors as a medical tourism destination (Medical Tourism Marketplace, n.d.-n).

Malta: With a healthcare system dating back to the 1300s, Malta has earned the MTI's 27th rank due to its well-oiled healthcare infrastructure and small size, which allows for convenient access to hospitals and facilities. With leading medical professionals, a low cost of care, surrounding tourism attractions, and common English skills, Malta's become a popular choice for orthopedic care, dental work, and esthetic treatments, though none of the island's hospitals are JCI accredited and Malta's distinct lack of public transport could affect tourists' journeys upon arrival (Medical Tourism Marketplace, n.d.-h).

Mexico: Located conveniently for North American patients, Mexico has been able to build up a robust medical tourism industry mainly patronized by American and Canadian citizens. Low travel costs and low cost of care have bolstered Mexico's popularity for travel for treatment, supplemented by the positive quality of treatment, high rates of English fluency among doctors, insurance coverage options, and a wonderful culture and climate for recovery. Mexico's wide variety of tourism destinations and pleasant climate also play a factor into Mexico's increasing popularity and 42nd MTI ranking, though a lack of standardization across facilities and a lack of medical liability are factors to consider ahead of booking treatment in Mexico (Medical Tourism Marketplace, n.d.-i).

Oman: Yet another one of the Persian Gulf's upcoming medical tourism hotspots, Oman has become the world's 13th top medical tourism destination according to the MTI. With a growing tourism economy and government investment to bring Omani medical tourism offerings up to par with their Middle Eastern peers, Oman's high-quality private sector healthcare is becoming increasingly known amongst international medical tourists.

Philippines: Another destination outfitted with pristine beaches, a unique culture and clear blue waters, the Philippines has become highly regarded for both its medical tourism and leisure tourism offerings. With extra support to the medical tourism industry granted by the country's government, the Philippines' Western-trained medical professionals and comprehensive malpractice policies have made the Philippines ranked 24th on the MTI, with many patients choosing this destination electing to turn their medical tourism journey into a full-blown vacation. That being said, the Philippines does not have the same quality of care as other nearby destinations, and reaching the island nation can take a considerable amount of time for patients flying in from North America and Europe (Medical Tourism Marketplace, n.d.-q).

Poland: Poland has similarly boosted its reputation in the global medical tourism market over recent years, earning the Medical Tourism Index's 29th spot. Mainly popular amongst nearby Europeans, Poland's cosmetic surgery and dental sectors have led its travel for treatment industry, with low costs, high-quality doctors, and high-standard medical facilities all contributing to the nation's success. However, a lack of options and minimal accreditation likely make Poland a more attractive option to fellow Europeans than to other international travelers (Medical Tourism Marketplace, n.d.-j).

Portugal: With a unique trisystem approach to healthcare, Portugal has become known for its high-quality standard medical care, dental work, reproductive services, and cosmetic surgeries performed by the country's medical tourism industry. Coupled with JCI accreditation, low costs, and tourism opportunities, Portugal has become the MTI's 21st top destination, though the Portuguese's leisurely approach to getting things done and poor road infrastructure may prove frustrating to medical tourists (Medical Tourism Marketplace, n.d.-k).

Saudi Arabia: Though less developed than its other Middle Eastern counterparts, major investment from the Saudi government is beginning to solidify Saudi Arabia's foothold in the international medical tourism industry, now sitting at the MTI's 35th spot. With plenty of tourism opportunities—though these may be offset by Saudi Arabia's cultural differences—anticipated further financial investment into Saudi Arabia's travel for treatment sector should continue growing the country's reputation amongst potential patients.

Singapore: The Medical Tourism Index's second-ranked medical tourism destination, Singapore's enduring commitment to cleanliness and excellence have brought Singapore's healthcare system to ultrahigh levels of quality, which has also been ranked by Bloomberg as the world's most efficient healthcare system. With most major hospitals backed by JCI accreditation and a widespread frequency of English fluency, Singapore's acclaimed healthcare system is only diminished in attractiveness to medical tourists by its difficult accessibility via flight from certain regions, as well as high cost compared to other Asian countries and uncomfortably humid climate, though Singapore's consistently impressive procedure results may render these other factors obsolete for travelers focused on quality above all (Medical Tourism Marketplace, n.d.-l).

South Africa: While South Africa's sprawling and diverse landscape offers major enticements to adventure-focused travelers, Africa's southernmost country has built its medical tourism industry up into the MTI's 14th best travel for treatment destination, particularly as many medical tourists pair their procedures with a safari excursion. Plastic surgery has become South Africa's main sector of procedures, with growing popularity in dentistry, heart, and obesity surgeries. Known for personalized, quality, and affordable care, South Africa has been limited by its lack of researchable facilities, a lack of implemented malpractice policies, and potential cultural and language barriers (Medical Tourism Marketplace, n.d.-m).

Spain: One of the world's best-known medical tourism destinations, Spain's dual-positive offerings of high-quality medical facilities and vacation spots have turned the European country into the MTI's 4th top medical tourism destination worldwide. Easy to get around and travel within, Spain's quality healthcare system has become an ideal choice for medical tourists looking to merge procedure with pleasure, though differing standards of care for nursing staff and high procedure prices in comparison to other medical tourism destinations must be taken into account (Medical Tourism Marketplace, n.d.-o).

Taiwan: Taiwan has become regarded as a "hidden gem" of the Asian medical tourism market, particularly in Southeast Asia. With advanced facilities highly ranked by the WHO, Taiwan has shirked accreditation from the JCI and instead elected to accredit its hospitals through its own Joint Commission of Taiwan. Specialist surgeries like cardiovascular, orthopedic, spinal surgeries, and organ transplants are all common procedures performed by Taiwan's medical tourism industry, though the country's tensions with China could present difficulty to medical tourists coming in from certain regions in Asia (Medtourpress, 2021).

Thailand: With over 1000 hospitals in the country, Thailand has become a major force in the international travel for treatment market thanks to its high cost-to-quality ratio and ample opportunity for tourism. With 37 JCI-accredited institutions, Thailand has taken the 17th rank on the Medical Tourism Index and has become a popular choice for plastic surgery procedures, dental work, fertility treatments, and orthopedic and cardiac surgeries. On the other hand, many of Thailand's health professionals only work part time, and the country has a distinct lack of emergency transport vehicles, which could play into a medical tourist's ultimate results (Medical Tourism Marketplace, n.d.-p).

Turkey: Consistently ranking in the world's top medical tourism destinations, Turkey has become an increasingly common choice for medical tourists hailing from both Europe and Asia due to its convenient location straddling the border of the two continents. With a number of natural wellness attractions like thermal springs, Turkey has been able to add medical tourism opportunities in plastic surgery, hair transplants, fertility treatments, stem cell transplants, and ocular care to its comprehensive offerings to travelers. JCI accreditation throughout many of Turkey's main hospitals helps bolster the country's reputation as a medical tourism hub, with competitive pricing and a welcoming culture attracting further medical tourists and nabbing Turkey the MTI's 30th spot. That being said, some Turkish hospitals are not as stringent with their medical records as Western facilities, and doctors oftentimes don't speak English, though many medical tourism agencies can help arrange a translator to help combat this issue (Medical Tourism Marketplace, n.d.-r).

United Kingdom: Rounding out the MTI's top medical tourism destinations is the United Kingdom, ranked at fifth place. While the United Kingdom has gained repute for its superb public healthcare system, a growing private sector has made the United Kingdom an increasingly popular destination for medical tourism, especially with its highly qualified medical professionals, state-of-the-art facilities, and use of advanced technologies and medical techniques, though the country's safeguarding of its medical tourism statistics has made it difficult to ascertain the nation's most popular procedures (Jagyasi, n.d.).

The United States: The United States, as one of the most medically and technologically advanced countries in the world, has continued to be a top medical tourism destination for global patients and US residents alike, offering both an impressive international and domestic travel for treatment sector. With a commonly spoken language and an internationally famous culture, plus inbound flights from most global destinations, the United States is expected to remain one of the strongest medical tourism locales moving forward—though its high and opaque pricing models could majorly put off patients in search of savings, including US citizens.

While these aforementioned nations are currently leading the modern medical tourism industry, numerous other countries—like Kuwait, Lebanon, Guatemala, Jamaica, Iran, Lithuania, Morocco, Panama, Qatar, Russia, Tunisia, Slovakia, and Vietnam—are becoming viable choices for travel for treatment purposes, and should be further explored by potential patients looking to understand all of their available options.

6.2 DIVING DEEPER INTO DUE DILIGENCE

After selecting the right medical tourism destination, it is imperative for patients to continue their search into selecting said country's best healthcare facility for their needs, including all the requisite accreditation confirmations and evaluating the facility according to the guidelines presented in Chapter 5. Other important factors to take into consideration are an institution's outcomes and success rates, level of follow-up care, the location of a treatment facility, pre- and postoperative travel and accommodation arrangements, a hospital's reputation both in the industry and within the local population, patient testimonies, and other general concerns surrounding the healthcare facility of choice.

Regarding surgeon selection, patients would be wise to ascertain the experience, knowledge, and credentials of the doctor they plan to perform their procedures with and should be encouraged to not always go with the cheapest option, despite hopes to take advantage of medical tourism's low costs. Medical tourists should focus their search on licensed, board-certified doctors, and should establish telecommunication or some form of communication with said surgeon ahead of travel. Agencies like MyMedChoices™ that specialize in medical tourism and travel for treatment can help provide patients assistance and advice as a facilitator and help guide tourists through the entire medical tourism process.

6.3 EXPLAINING MEDICAL TOURISM'S COST DIFFERENTIAL

While travel for treatment's cost-saving potentials may excite many medical tourism aspirants, patients should be aware of just *why* procedures can vary so greatly in cost from country to country. Particularly in developing nations, lower cost of human labor, lower real estate values, and low construction costs have helped drive down the price of medical tourism procedures. Likewise, efforts from national governments to amplify their internal medical tourism industry has resulted in subsidies and lowered taxes on the industries, while medical supplies, equipment, and medications can also be acquired at a fraction of the cost to their value in countries like the United States. On that same note, lower administrative costs and documentation expenses and lower costs of legal proceedings also play an important factor in many medical tourism destination's low pricing structure.

6.4 UNDERSTANDING THE MENTAL EFFECTS OF MAJOR SURGERY

Prepare as a patient may for any and all circumstances that come their way, one silent menace has the potential to negatively affect even the healthiest person's surgery results: postsurgical depression. Though an increasingly common side effect of surgical procedures, many patients remain unfortunately ignorant about their propensity to become depressed during the recovery process. As doctors often fail to warn their patients about this condition before it's already too late. While consulting with a physician, exercising, eating well, and maintaining a positive outlook can help patients work through their postsurgical depression, patients should be keen to keep a close eye on their mental health throughout their medical tourism process to help stop depression in its tracks, as well as understand the realities of their possible postsurgical side effects before setting off on their medical tourism travels, as securing mental health and a positive state of mind should be the very first step in the journey to wellness.

KEYWORDS

- **medical tourism**
- **travel for treatment**
- **mental health**
- **medical tourism destinations**
- **due diligence**

REFERENCES

Abu Zaid, M. *Egypt Launches Medical Tourism Program for Foreign Patients*. Arab News, November 3, 2021. https://www.arabnews.com/node/1960846/middle-east

Budiman, A. *Key Findings about U.S. Immigrants*. Pew Research Center, September 22, 2020. https://www.pewresearch.org/fact-tank/2020/08/20/key-findings-about-u-s-immigrants/

Center for Disease Control and Prevention. *Medical Tourism—Chapter 9—2020 Yellow Book*, June 24, 2019. https://wwwnc.cdc.gov/travel/yellowbook/2020/travel-for-work-other-reasons/medical-tourism

Find Private Clinics Canada. *Medical Tourism in Canada; Guide to Medical Treatment in Canada*, 2022. https://www.findprivateclinics.ca/resources/general/medical-tourism.php

Gulf News. *Plastic Surgery among Top Five Procedures of Health Tourism in Dubai*, February 18, 2018. https://gulfnews.com/uae/health/plastic-surgery-among-top-five-procedures-of-health-tourism-in-dubai-1.2175202

HTI Conference. *Health Tourism in Italy Has a Positive Outlook*, November 14, 2017. https://hticonference.com/health-tourism-italy-positive-outlook/

IMTJ. *Plans for Reversing Medical Travel from the Bahamas*. Laing Buisson News, March 31, 2021. https://www.laingbuissonnews.com/imtj/news-imtj/plans-for-reversing-medical-travel-from-the-bahamas/

IMTJ. Dominican Republic Develops Long Term Plan for Medical Tourism. *Laing Buisson News*, March 18, 2022. https://www.laingbuissonnews.com/imtj/news-imtj/dominican-republic-develops-long-term-plan-for-medical-tourism/

Jagyasi, P. *Guide to Medical Tourism in UK, Important Facts and Features*; Dr. Prem, n.d. https://drprem.com/medical-tourism/medical-tourism-in-uk

Leggat, P. *Medical Tourism*; Australian Family Physician. January 2015. https://www.racgp.org.au/afp/2015/january-february/medical-tourism

Matheson. *The Pros and Cons of Medical Tourism*; Lexology, February 13, 2009. https://www.lexology.com/library/detail.aspx?g=66db0ce5-f789-4677-9699-07d0660cda79

Medical Tourism Association. *The Medical Tourism Index 2020–21*, November 22, 2021. https://www.medicaltourism.com/mti/home

Medical Tourism Magazine. *Bahrain Leads the Way for Medical Tourism in the GCC*; Medical Tourism Magazine, December 16, 2021a. https://www.magazine.medicaltourism.com/article/bahrain-leads-the-way-for-medical-tourism-in-the-gcc

Medical Tourism Magazine. *Armenian Medical Tourism Strategies to Get Overdue Checkup*; Medical Tourism Magazine, December 16, 2021b. https://www.magazine.medicaltourism.com/article/armenian-medical-tourism-strategies-to-get-overdue-checkup

Medical Tourism Magazine. *Japan—Realizing Medical Tourism Potential*; Medical Tourism Magazine, December 16, 2021c. https://www.magazine.medicaltourism.com/article/japan-realizing-medical-tourism-potential

Medical Tourism Marketplace. *Medical Tourism to Austria*; Medical Tourism Marketplace, n.d.-a. https://www.health-tourism.com/medical-tourism-austria/

Medical Tourism Marketplace. *Medical Tourism to Colombia*; Medical Tourism Marketplace, n.d.-b. https://www.health-tourism.com/medical-tourism-colombia/

Medical Tourism Marketplace. *Medical Tourism to Costa Rica*; Medical Tourism Marketplace, n.d.-c. https://www.health-tourism.com/medical-tourism-costa-rica/

Medical Tourism Marketplace. *Medical Tourism to Greece*; Medical Tourism Marketplace, n.d.-d. https://www.health-tourism.com/medical-tourism-greece/

Medical Tourism Marketplace. *Medical Tourism to Hungary*; Medical Tourism Marketplace, n.d.-e. https://www.health-tourism.com/medical-tourism-hungary/

Medical Tourism Marketplace. *Medical Tourism to India*; Medical Tourism Marketplace, n.d.-f. https://www.health-tourism.com/medical-tourism-india/
Medical Tourism Marketplace. *Medical Tourism to Jordan*; Medical Tourism Marketplace, n.d.-g. https://www.health-tourism.com/medical-tourism-jordan/
Medical Tourism Marketplace. *Medical Tourism to Malta*; Medical Tourism Marketplace, n.d.-h. https://www.health-tourism.com/medical-tourism-malta/
Medical Tourism Marketplace. *Medical Tourism to Mexico*; Medical Tourism Marketplace, n.d.-i. https://www.health-tourism.com/medical-tourism-mexico/
Medical Tourism Marketplace. *Medical Tourism to Poland*; Medical Tourism Marketplace, n.d.-j. https://www.health-tourism.com/medical-tourism-poland/
Medical Tourism Marketplace. *Medical Tourism to Portugal*; Medical Tourism Marketplace, n.d.-k. https://www.health-tourism.com/medical-tourism-portugal/
Medical Tourism Marketplace. *Medical Tourism to Singapore*; Medical Tourism Marketplace, n.d.-l https://www.health-tourism.com/medical-tourism-singapore/
Medical Tourism Marketplace. *Medical Tourism to South Africa*; Medical Tourism Marketplace, n.d.-m https://www.health-tourism.com/medical-tourism-south-africa/
Medical Tourism Marketplace. *Medical Tourism to South Korea*; Medical Tourism Marketplace, n.d.-n. https://www.health-tourism.com/medical-tourism-south-korea/
Medical Tourism Marketplace. *Medical Tourism to Spain*; Medical Tourism Marketplace, n.d.-o. https://www.health-tourism.com/medical-tourism-spain/
Medical Tourism Marketplace. *Medical Tourism to Thailand*; Medical Tourism Marketplace, n.d.-p. https://www.health-tourism.com/medical-tourism-thailand/
Medical Tourism Marketplace. *Medical Tourism to the Philippines*; Medical Tourism Marketplace, n.d.-q. https://www.health-tourism.com/medical-tourism-philippines/
Medical Tourism Marketplace. *Medical Tourism to Turkey*; Medical Tourism Marketplace, n.d.-r. https://www.health-tourism.com/medical-tourism-turkey/
Medical Travel Quality Alliance. *Dominican Republic Dangerous for Medical Tourists, Risks too High in India, Mexico, Thailand, South Korea: New MTQUA Report*, December 21, 2015. https://mtqua.org/dominican-republic-dangerous/
Medtourpress. *Why Medical Tourism in Taiwan?* February 21, 2021. Medtourpress. https://medtourpress.ir/en/medical-tourism-taiwan/
Pacific Prime. *Medical Tourism: What You Need to Know before You Go*, March 22, 2021. https://www.pacificprime.com/blog/medical-tourism-what-you-need-to-know-before-you-go.html
Pierson, B. *Unproven Stem Cell Therapy Subject to FDA Regulation—11th Circ.* Reuters, June 2, 2021. https://www.reuters.com/legal/litigation/unproven-stem-cell-therapy-subject-fda-regulation-11th-circ-2021-06-02/
Royal Medical. *Excellent Czech Medicine*, July 13, 2021. https://www.royal-medical.eu/about-us/excellent-czech-medicine/
Solomon, B. *10 Essentials You Need to Pack for Your Medical Tourism Trip*. Medical Tourism Magazine, December 16, 2021. https://www.magazine.medicaltourism.com/article/10-essentials-need-pack-medical-tourism-trip
Thopson, S. *Medical Tourism in Argentina—A Source of Revenues and Jobs*. Tourism News for Travel and Tourism Industry Professionals, March 22, 2022. https://www.tourism-review.com/medical-tourism-in-argentina-growing-fast-news5446
Vaidam. *Medical Tourism in Egypt*. For Medical Travelers, January 29, 2022. https://www.vaidam.com/knowledge-center/medical-tourism-egypt
Whitlock, J. *Why Patients Are Turning to Medical Tourism*. Verywell Health, September 11, 2021. https://www.verywellhealth.com/understanding-medical-tourism-4069869

CHAPTER 7

THE NEXT STEPS IN RESETTING HEALTHCARE

ABSTRACT

Though national healthcare systems and private insurance providers have long gate kept their patients' access to different care choices, international medical patients are becoming increasingly aware of their ability to travel and acquire high quality care on their own accord. This chapter explores the trend toward value-based care, the current status of the US healthcare system, the debate between healthcare as a right or as a commodity, and how both patients and doctors alike can make the most of the modern medical industry—especially post-pandemic.

When the majority of consumer's lifestyles—where we live, where we eat, what we wear, who we talk to—are determined by individual tastes and choices, the general population has not been offered the same opportunity to hand-select their own medical care—until now. With health deeply ingrained into the mind of the global population after the COVID-19 pandemic, the present climate has become a ripe opportunity for consumers across the world to reassess their relationship with healthcare and catalyze innovation in the industry toward patient desires. While contemporary healthcare, particularly within the United States, has gained a bad rap over the years from a lack of transparent pricing, excessive costs, and a lack of flexibility, postpandemic consumers are now in the unique and singular position to reshape the international healthcare marketplace.

Rather than be restricted by the confines of a country's internal healthcare systems, consumers can put autonomy and self-determination back on the agenda through personalizing and customizing their healthcare choices,

Medical, Dental, and Wellness Tourism: A Postpandemic Perspective. Mary Schreiber Swenson, PhD, and Amit Bansal (Authors).

particularly by harnessing the benefits of travel for treatment and medical tourism. As a person's health, and by extension, their family's health, plays one of the most important factors in quality of life, empowering global consumers to make their own bespoke healthcare decisions could potentially change the international community's overall quality of life in turn.

With this in mind, there seems ample opportunity to offer value-based healthcare to the international community, particularly as medical tourism and travel for treatment become increasingly common. Under this healthcare model, patients pay healthcare providers (whether hospitals, doctors, or technicians) for the direct value of the results of the services they provide, rather than a flat-fee or capitation approach. Considering health is certainly not a one-size-fits-all kind of industry, value-based healthcare argues that services and their pricing structure should similarly reflect the uniqueness of each individual case while aiming to enhance patient experience, reduce cost of healthcare, and improve population health along the way.

For consumers, value-based healthcare offers the chance for patients to directly tackle and treat injuries or illnesses at their start and achieve better outcomes, allowing patients to curtail the development of chronic disease and spend less money to reach better levels of health. By stopping potentially life-changing problems in their tracks, patients can avoid expensive additional doctor's visits and extraneous prescription medications, securing themselves better short and long-term health in the process.

Though value-based healthcare undeniably benefits the consumer, the positive effects of value-based healthcare extend to all parties involved, including providers. By focusing on preventative care, engagement with patients, and quality care, providers can cut down on the considerable quantities of time needed to treat and manage chronic diseases, and minimize the risk associated with capitated payments, even extending the financial earnings of for-profit providers under value-based healthcare. Payers similarly benefit by controlling costs and creating a more diverse patient pool and can boost their advantages by creating value-based healthcare payment bundles adjusted directly to a patient's specific health situation.

On the supplier's side of things, suppliers can benefit from the positive optics of their products and services producing positive results and satisfied patients. With national health expenditures on the rise, value-based healthcare is anticipated to be a major selling point to consumers moving forward—plus, further steps forward toward value-based healthcare could have a ripple effect among prescription medication manufacturers, encouraging these manufacturers to offer these drugs and prescriptions at their *actual value*, rather than inflated for-profit prices, to consumers. Society as a whole benefits, too, as the population becomes healthier overall while similarly reducing their healthcare spending, a particularly pertinent factor for countries with a high annual healthcare expenditure such as the United States to consider.

Within value-based healthcare are a number of delivery models shifting the dynamics between patients and providers, increasing focus on interteam coordination and an open sharing of patient data among staffers that allows for more precise care and quantifiable results. Two models growing in popularity are medical homes and accountable care organizations, which each offer their own benefits under a value-based healthcare system.

One such model making waves throughout value-based healthcare is patient-centered medical homes (PCMHs), which integrate specialty, primary and acute care into a coordinated, team-oriented approach to treatment. Here, a patient's primary physician helps direct all of the individual professionals involved in a patient's routine care for a more comprehensive care plan. Rather than take place in a singular physical facility, the term PCMH refers to the collaborative nature of treatment, particularly as this value-based healthcare model relies extensively on the sharing of electronic medical records among all team members involved for increased communication and a more robust understanding of a patient's specific needs. With this focus on data sharing, doctors are able to strategically pinpoint and target the next steps of care by using test results and other procedures performed by other members of a patient's team, while patients themselves are able to cut down on their costs by receiving more efficient care.

Accountable care organizations, or ACOs, help provide value-based healthcare to Americans covered by Medicare, as designed by the Centers for Medicare and Medicaid Services. This, much like PCMH, is a patient-oriented structure that relies on a team of coordinated medical professionals to ramp up accuracy and efficiency, with all team members sharing the same risks and rewards—both financial and patient healthwise—to continue driving down cost while maintaining quality of treatment. Rather be encouraged to continue ordering tests and other extraneous treatments to drive up pay like doctors in fee-for-service models, the utilization of ACOs' patient-oriented approach has shown demonstrable "improvements in outcomes such as hospital readmissions, adverse events, patient engagement, and population health" (NEJM Catalyst, 2017).

It is no secret that the healthcare industry is in need of reform, but moving from the traditional flat-fee model to a value-based healthcare system will not be a simple transition. In order for the healthcare system to function effectively, all stakeholders—including patients, doctors, and insurance companies—must work together toward common goals of improving the population's health and lowering healthcare expenditures, especially as implementing a value-based healthcare system continues to be highly regarded as the best course of action to achieve these objectives.

7.1 A RIGHT OR A COMMODITY?

There has been a lot of controversy over the years about whether healthcare is a commodity or a right. The definition of healthcare is difficult to nail down, as it can be interpreted in many ways by taking into account near-countless factors. While some people believe that everyone should have access to quality healthcare as a right, and others think that it should only be available to those who can afford it as a commodity, the debate is sure to continue for many years to come (Spence, 2018).

That being said, healthcare *technically* is an internationally recognized human right according to the United Nations Universal Declaration of Human Rights, which was signed into effect in 1948 by the United States and 47 other countries from across the world. According to article 25 of the declaration, "everyone has the right to a standard of living adequate for the health and well-being of himself and of his family, including food, clothing, housing and medical care and necessary social services, and the right to security in the event of unemployment, sickness, disability, widowhood, old age or other lack of livelihood in circumstances beyond his control" (United Nations, n.d.).

Despite the United Nations' aspirational declaration, healthcare is very clearly *not* regarded as a human right across the international economy, particularly as one in every three people in the world does not have access to safe drinking water—one of the largest contributing factors to poor health in the populations of developing countries, alongside a lack of accessible sanitary resources (World Health Organization, 2019). If healthcare was indeed a right, it would be the moral obligation of all international governments to make medical care both affordable and attainable for all; while some countries have indeed adopted generous public healthcare systems, others, like the United States, continually position healthcare as a commodity rather than a right.

However, various policies across US presidential administrations have continued to muddy the waters on healthcare's true role. Signed into law under President Barack Obama, the US Patient Protection and Affordable Care Act oddly defined healthcare as both a commodity and as a right within the very same bill, further confounding the definition of modern healthcare—an issue that could be simply resolved by national governments throughout the public healthcare sector.

While private sector care plays a vital role in the functionality of the international healthcare economy, government decisions relating to public healthcare infrastructure unduly impact the healthcare opportunities for a country's lower income population, especially when considering a nation's higher net worth residents that can easily afford private care. With this in mind, progressive government regulations on healthcare and close negotiations with the private sector could help create a more equitable playing ground for healthcare for all, though the influence of lobbies and major healthcare providers on government officials have oftentimes put the financial viability of for-profit healthcare companies on priority over the well-being of the entire US population.

To help equalize the playing field, the United States has implemented a number of policies aiming at reducing kickbacks—or payment incentives for patients to patronize a certain doctor, provider, or hospital—to ramp up patient distribution. However, the United States' Anti-Kickback Statute dictates that any medical provider that uses kickback incentives to promote their services is committing fraud, even when a patient is allowed to pay for said services with public insurance options like Medicaid and Medicare, among others. Though this kickback-focused legislation can indeed help reduce the exploitation of the patient's well-being for the financial benefit of the other parties involved in the kickback scheme, the policy has the unfortunate side effect of reducing free referrals and important information acquisition among medical professionals, causing some doctors to instead accept roles at hospitals in exchange for their preestablished customer base—something that, depending on the definition, could also be defined as a kickback in its own right (Bush and Baker, 2014). According to Jonathan Bush and Stephen L. Baker, authors of *Where Does It Hurt?: An Entrepreneur's Guide to Fixing Health Care*, cutting back on the United States' anti-kickback laws could help promote more equitable healthcare distribution, while still allowing medical professionals and providers to be financially incentivized to provide care to Americans that rely on public insurance options.

There is likewise been a widespread call from the US doctors for tort reform, which would put in place limitations for medical malpractice lawsuits and leave doctors with less liability, as well as driving down their sky-high insurance premiums—which, naturally, are pushed down to the patient in the form of service price hikes. This has become a point of contention between patients and doctors, as patients enjoy tort provisions as a form of protection in the face of a procedure gone-wrong, while medical professionals oftentimes dislike tort policies due to concerns about patients abusing malpractice policies for the sake of financial gain. As medical data records become increasingly digitized, however, a wealth of readily available medical information about a patient's medical past could help resolve any doubts—the culprit of many medical malpractice lawsuits—along the way of the treatment process, as well as open up the door for further data-driven industry innovations.

Within the United States, the US government plays a singular role in the protection and privacy of its residents' data via its legislative choices; though, according to Bush and Baker, the state's stance on date security has altogether limited the opportunity for a robust healthcare data market, where parties would normally share relevant or requested information in exchange for monetary compensation. An open healthcare data market that acknowledges data as a source of revenue in the United States would further prompt healthcare providers and professionals to provide efficient service and continue acquiring data across all levels of their practice and could be the precursor to a bustling international data exchange that could positively impact medical tourists and travel for treatment candidates through streamlined doctor-to-doctor communication unburdened by the borders on the map.

Further research into the potentials and efficiencies of electronic health record (EHR) applications could see a larger push into reforming medical data laws in the United States, as recent studies have demonstrated the implementation of EHRs' positive impact on medical results and the healthcare system as a whole while still maintaining an important perspective on patient's comfortable confidentiality levels.

> The promising EHR implementation systems, people, process, and product factors play an integral role in the fate of its implementation. The stakeholder's benefit from the systems which protect the patient's needs and ensure their privacy. The access to accurate and complete clinical information is the main component of effective decision-making. This is facilitated by decision-support EHR tools—BPAs—and designed for behavioral health integration with the needs of healthcare institutions and the benefits of improving the patient experiences, for instance, alcohol use. The system can be used to update the current condition of a patient as input to obtain

> a corresponding recommendation for medical tests, possible diseases, and treatment plans. Research indicated that the EHR "active choice" significantly increased influenza vaccination rates and ordering of colonoscopy and mammography screening services….
>
> ….The shift in the use of EHR by health builds trust and presents an opportunity to monitor admission, diagnosis, and outcome to inform public health policy and service provision. (Adane et al., 2019)

While necessary to be implemented on a country-by-country basis, the emergence of EHR within the medical community, and its acceptance by the general population, could later see this important medical data shared between international health professionals and institutions through a worldwide EHR application. Through this course of action, doctors across the world—even in the most remote markets—could positively benefit from having access to the most up to date and relevant medical information on the market, which has the trickle-down effect of doctor's subsequently providing patients with more thorough and improved-upon care. Consumers, too, now incensed by the COVID-19 pandemic to invest more money into their personal wellbeing could similarly be enticed in by a well-designed application that gives them unbridled access to their own medical records, making the potentials of EHR implementation ripe for all parties involved.

However, that is not to say that opening up the data market for EHR applications would be without its challenges; alongside the introduction of a new lucrative market comes new opportunities for its abuse, particularly as medical data and patient privacy are put on the chopping block in the case of EHRs. Still, preestablished legislation—like the 1996 Health Insurance Portability and Accountability Act—should provide protection to consumers and their confidentiality needs, and additional legislation could be drafted in turn to resolve any issues that pop up along the way post-implementation.

Though the present state of the healthcare system clearly requires government reform and intervention, the state's role in the modern healthcare system should continue to place business as the key middleman between the government and the patient as the facilitator of care. Unfortunately, the design of the current public US healthcare options—including Medicare Advantage, which gives a patient's care to a private insurer to be performed underneath a capitated amount—has suffered underneath countless hurdles and red tape brought forth by varying state legislation, causing many US states to regress as close as possible back to a for-profit privatized healthcare model. With a little bit of federal government intervention, Medicare and

Medicare Advantage could be reformed to allow patients to truly thrive under a true capitated healthcare model, as well as giving providers savings and an extended pool of professionals to choose from as they attempt to deliver the best care available to their publicly insured clients.

On that same note, many US states have made their healthcare process overly bureaucratic, putting in place limitations and restrictions that have a much larger negative blowback on patients than it does on providers. Licensing has become a major point of contention in state-run healthcare, as lobbyists run rampant to keep regulations in place that maintain the supply of healthcare professionals at a consistent level, as a means to protect preestablished businesses from the entrance of new competitors into the market. This means a lesser availability of qualified healthcare providers for patients to choose from, a practice only bolstered by medical schools' own influence from lobbyists, as the learning facilities personally limit the supply of how many doctors are trained in a specialty every year.

Many states have limited the number of new medical businesses or start-ups allowed to operate within the state to purportedly stop oversupply, preventing consumers from dictating the market with their own spending power and altogether putting the fate of the healthcare system in the hands of lobbyists. Likewise, the rigorous licensing and credential standards lobbied for in some states have forced out well-equipped staffers, like orderlies, from performing tasks well within their realm of experience at a lower cost than higher credentialed professionals, and similarly keep the pool of talent reliant on membership to certain guilds, oftentimes under the guise of safety and patient protection.

That is not to say that accreditation and licensing should be done away with all together; many specialties, like anesthesiology, require the strict adherence to certain rules and regulations to maintain the health and livelihood of the patient they serve and will continue to benefit from licensing standards. However, as advocated by Bush and Baker, the world of healthcare could benefit from more open licensing standards, which would allow medical professionals to establish their value based on proof of work and tangible results, rather than a piece of paper as it is now. This in turn would create increased competition, driving down prices and allowing patients greater choice in selecting a quality healthcare provider that fits their needs on their own terms.

The US healthcare system could further benefit from government intervention in allowing residents to find and acquire insurance options in whatever state they choose, though this is currently prohibited by law after rigorous lobbying. Allowing patients to fluidly move across states and control the power of their spending dollars would open up competition across the country and create a fairer, more transparent marketplace—precisely what healthcare lobbies do *not* want to happen—and these effects could positively ripple throughout the domestic travel for treatment sector, making internal medical tourism more attainable for all.

In the grander scheme of things, the small pool of mega-insurance carriers gives little wiggle room or personal choice for modern consumers to curate their own healthcare. Instead, a number of preestablished packages that eerily encompass near-identical components across providers keeps patients trapped within said providers' carefully constructed boxes without input. While start-ups may be willing to experiment with new programs or structures, aforementioned limitations set by state governments on start-up businesses have left the mega-insurance carrier programs at the top of the food chain despite their lack in variety. Plus, new insurance carriers assume a risk far beyond that of their larger competition; less patients means more money necessary to cover their payouts, and subsequently, more money to cover this risk, though this trend begins to dissipate around the million-patient threshold. This provides an intriguing opportunity for federal government intervention in the healthcare industry, much like it does through its federally owned mortgage companies for the real estate sector, which could help provide funding to insurance startups as they grow into becoming viable competition in the overall healthcare marketplace, and further balance the scales between healthcare as both a commodity and a right.

7.2 HOW PATIENTS AND DOCTORS CAN MAKE THE MOST OUT OF MODERN HEALTHCARE

Though the government and lobbyists have dictated the rules of the contemporary healthcare system, it is up to patients and doctors alike to figure out how to make the best of their role within the system for the good of themselves and for others. While policies are slow to change and reforms take time to make headway, integral players in the system like healthcare professionals and consumers can begin working toward systemic change themselves through the course of their own individual actions.

For patients, one of the most important steps they can change toward changing the playing field is simply talking with one another. While talking about finances and how much things cost has become taboo over the years, creating an open and honest discourse about value of things almost always helps work toward equity—just look at new advances in the US workplace as an example, where increasingly open discussions about pay levels have thrust labor exploitation in the spotlight and created a ripe opportunity for workers to demand fair pay from their employers (Conlan, n.d.).

Within the US healthcare system, many patients are not privy to the cost of the treatment or service received until they receive the bill in the mail. This has allowed providers to set arbitrary prices that can vary from person to person or facility to facility, driving up low-cost pharmaceutical products to exorbitant price tags, much like the controversial $629 bandage bestowed upon a one-year-old girl back in 2016, and reports of hospitals price gouging generic medications like aspirin to 100-fold its value upon billing (Kliff, 2016). Though standard items driven to crazy prices may grab headlines, it is where patients lack knowledge about pricing that the costs truly begin to soar, as the average patient would not be able to accurately assess all of the costs associated with a complex surgery—and as such, hospitals can dictate prices as they wish, forcing consumers who wish to live to see another day to pay the facility's designated fee anyway, regardless of what is fair or not. By removing the social stigma from openly discussing healthcare pricing *en masse*, newly enlightened consumers may be able to gain a better perspective on the true costs of care, as well as band their voices together to fight for a more transparent and fairly priced healthcare industry.

Plenty of this can be accomplished by individual patients setting out to learn more about their afflictions and sharing them with the world, whether in-person or throughout online communities. The popularization of digital media and virtual communication presents a particularly potent area for patients to learn and grow together, as forums and chat rooms give the chance for people to share their symptoms and experiences outside of the hospital setting, and act as mentors or advisors on what facility to visit and what doctor to receive surgery from. Patients' intent on seizing back the fate of their own healthcare opportunities should likewise begin requesting their medical data from their providers, particularly as many providers claim ownership over these records and keep them away from patients until otherwise requested. With information acting as the great equalizer for patients in the modern-day healthcare system, these online communities are creating a consumer-driven healthcare market on their own that relies on patient experience, rather than provider statistics, as a point of reference.

Doctors are put into a unique position as the middle-point between patients and providers, needing to play to both parties to remain in good repute throughout the medical community. That being said, doctors should focus on delegating any assignable tasks to other competent staffers and utilizing state-of-the-art technology to fill in the gaps, so that doctors can focus on what they do best: caring for patients within their specialty.

Of course, as technology continues to play an increasingly important role in medical care, even to the point of the emergence of so-called medical factories that perform specialty procedures like knee replacements via machines, doctors will need to lean into the person-to-person aspect of their careers, acting as a trusted advisor in the midst of machine dominance. Though doctors will undeniably still play an important role leading up to factory-produced procedures, establishing a real, human connection with patients will be more and more important for doctors as medical technology develops in turn.

This can extend into a doctor's practices, as well; rather than just handle diagnosis and treatment, doctors looking to make the most out of the healthcare system will begin offering patients care from point A to Z. From surgery to recovery to rehab, providing comprehensive care packages gives patients the best quality of treatment possible, while doctors retain patient loyalty and evolve the scope of their services to benefit from all ends of the treatment process. Likewise, as doctors will likely scope out the specifications and efficiency of different factories to choose what they wish to provide for their patients, doctors may act as the first tastemakers of sorts in the new era of the healthcare industry and should provide more nuanced selections than profit-forward major healthcare providers.

Doctors, much like patients, should also consider their power in numbers and organize among one another rather than subscribe to employment at a major hospital—though, given the major checks offered by these facilities, it is not difficult to see why many doctors go this route over private practice. However, hospitals are so consumed with filling quotas that they regularly diminish quality of patient care in favor of stacking numbers, and use doctors as a means to attract more patients into their doors. By teaming up and consolidating power with one another in private practice, doctors can provide an extremely high level of excellence to their patients and nurture a financially flourishing business as a result, a far cry from the poorly run and low-quality business practices of many modern hospitals, which can be bolstered to new heights of success through rigorous data acquisition.

At the end of the day, doctors and patients have the capabilities to begin reshifting the power dynamics of the modern healthcare system through their own hands. Though the state of healthcare may be intimidating—both in the

United States and on an international scale—a further emphasis on quality, accessibility, and fair pricing in tandem by patients and doctors could help push forward the postpandemic healthcare reset the world so desperately needs.

KEYWORDS

- **medical tourism**
- **travel for treatment**
- **value-based care**
- **accessibility**
- **healthcare**

REFERENCES

Adane, K.; Gizachew, M.; Kendie, S. The Role of Medical Data in Efficient Patient Care Delivery: A Review. *Risk Manage. Healthcare Policy* **2019,** *12*, 67–73. https://doi.org/10.2147/rmhp.s179259

Bush, J.; Baker, S. *Where Does It Hurt?: An Entrepreneur's Guide to Fixing Health Care*; Penguin, 2014.

Conlan, C. *The Truth about Discussing Your Salary*; Monster Career Advice, n.d. https://www.monster.com/career-advice/article/truth-about-discussing-your-salary

Kliff, S. The Case of the $629 Band-Aid. *Vox*, May 13, 2016. https://www.vox.com/2016/5/13/11606760/emergency-facility-fees-american-health-care

NEJM Catalyst. *What Is Value-Based Healthcare?* January 1, 2017. https://catalyst.nejm.org/doi/full/10.1056/CAT.17.0558

Spence, P. *Health Care Is Not a Service Commodity*; KevinMD.com, May 7, 2018. https://www.kevinmd.com/2018/05/health-care-is-not-a-service-commodity.html

United Nations. *Universal Declaration of Human Rights*, n.d. https://www.un.org/en/about-us/universal-declaration-of-human-rights

World Health Organization. *1 in 3 People Globally Do Not Have Access to Safe Drinking Water—UNICEF, WHO*, June 18, 2019. https://www.who.int/news/item/18-06-2019-1-in-3-people-globally-do-not-have-access-to-safe-drinking-water-unicef-who

CHAPTER 8

HEALTH, MENTAL WELLNESS, AND SEEKING CLARITY POST-COVID-19 THROUGH RESORT TRAVEL

ABSTRACT

The COVID-19 pandemic caused personal health and wellness to become one of consumer's core priorities, a trend that's expected to continue as the long-term mental and physical effects of the virus become more evident with time. As a result, consumers are seeking ways to bounce back both their minds and bodies from COVID-19 through wellness travel; this chapter analyzes pre and post-pandemic wellness travel trends, and how clarity has become a focus of luxury travel for treatment experiences.

Clarity has become an increasingly important focus of wellness for many consumers following the COVID-19 pandemic. In the wake of the most widespread global health crisis in a century, many international consumers—particularly young adults—have become progressively interested in ensuring that their environment is uplifting and positive toward their mental health; this has led to a growing demand for clarity in all aspects of life, from the food we eat to the air we breathe, but above all, in the mind.

This has become particularly evident in the rise of clarity-centric wellness retreats in recent years, even the years prior to the pandemic, with many resorts demonstrating a prioritization of their clients' mental clarity through amenities and offerings like sound healing, meditation, nutrition, and exercise opportunities like yoga. It's not difficult to see why these ventures have been steadily growing in popularity, particularly amongst medical tourists and travelers for treatment; clarity can bring improved mental focus and health, reduced stress levels, improved physical health, and increased productivity—all

Medical, Dental, and Wellness Tourism: A Postpandemic Perspective. Mary Schreiber Swenson, PhD, and Amit Bansal (Authors).

highly desirable results after the massive mental health challenges brought on by COVID-19 itself as a virus, and the extrapolating effects of continued isolation, the suffocation of being locked down in close quarters due to social distancing restrictions, governments' varying COVID-19 policies across the world, and, of course, massive economic upheaval.

> During the pandemic, about 4 in 10 adults in the U.S. have reported symptoms of anxiety or depressive disorder, a share that has been largely consistent, up from one in ten adults who reported these symptoms from January to June 2019. A KFF Health Tracking Poll from July 2020 also found that many adults are reporting specific negative impacts on their mental health and well-being, such as difficulty sleeping (36%) or eating (32%), increases in alcohol consumption or substance use (12%), and worsening chronic conditions (12%), due to worry and stress over the coronavirus. (Kaiser Family Foundation, 2021)

These issues have disproportionately affected the young adult population aged between 18 and 24, one of the most vulnerable populations to mental health disturbances due to the fact that their brains will not have fully formed until reaching the age of 25. In fact, more than half of the US young adult population, or 56%, experienced symptoms related to anxiety or depression throughout this time period, as well as more than double the amount of suicidal thoughts and substance abuse (clear signs of mental distress) than the statistics for the overall US adult population.

This can partially be attributed to continued economic hardships brought on by the pandemic, which directly correlate to a person's mental health status (American Psychological Association, 2005). According to the US Bureau of Labor Statistics (2021), 99 million US citizens eligible and able to work were unemployed in July 2020, a direct result of widespread layoffs and business closures brought on by COVID-19—and representing more than a quarter of the total population of the United States. This unduly affected young adults once more, a segment of the population usually the first to go in layoffs due to a lack of seniority in the workplace and their prevalence in the service industry, which, as previously explored, experienced widespread hurdles and shutdowns throughout the pandemic, a trend likewise reflected on an international scale.

Many young adults work in the sectors hardest hit by the pandemic—such as the service industry and manufacturing—often on part-time or temporary contracts with limited job protection. The informal sector, where almost 80% of the world's young workers are employed, was particularly impacted.

Altogether, the number of young people who are not in employment, education or training (NEET), already at 21% in early 2020, is likely to rise in the coming year....

> ....Young adults' employment prospects were being challenged by automation, as well as by disruption from the Fourth Industrial Revolution, before interrupted education opportunities and job losses set them further behind. Youth unemployment may increase across regions, given that more sectoral restructuring and shifting consumer habits are expected to trigger mass layoffs. Low-wage jobs—which could provide a safety net for young workers starting their careers—are also projected to decrease. (World Economic Forum, 2021)

Dubbed as "pandemials," a portmanteau of pandemic and Millennials, this global population is expected to have enduring financial issues for the rest of their lives due to the unfortunate card dealt to them throughout the pandemic, especially when coupled with their adolescence at the time of the 2008 global financial crisis; "for young workers, one month being unemployed at age 18–20 can cause a permanent income loss of 2% in the future" (World Economic Forum, 2021). Accounting for 15.5% of the global population, these 1.21 billion young adults across the world—who are supposed to be the backbone of the future's economy and workforce—are now at risk of becoming another "lost generation," unless robust action against this trend is taken—and fast (United Nations, 2020).

With the choices made in the years following the pandemic vital to the continued vitality of their brains—and subsequently, the quality of the rest of their lives—many young adults understandably are attempting to turn back the hands of time by making personal clarity and wellbeing a major priority postpandemic.

8.1 PRE-COVID WELLNESS RETREAT TRENDS

Long before the arrival of the COVID-19, millennials and young adults have been trending toward wellness retreats and holistic experiences with the intention of improving their mental health and overall personal clarity, seeking out transformational moments that later reverberate throughout the rest of their lives. Rather than singularly focusing on the vacation-heavy aspect of the traditional retreat, more and more prepandemic resorts began offering their visitors a wealth of wellness-related amenities and opportunities alongside the requisite leisure aspects associated with this form of travel.

According to the Global Wellness Summit's 2018 conference, prepandemic Millennials, incensed by their on-demand access via social media and the internet to wellness influencers and a wealth of available information, no longer are satisfied by the simple, spa-esque experiences the resorts of yesteryear have become known for (Dimon, 2018). As of 2018,

> As burned-out Millennials look to vacations to reset their health habits, wellness travel has swelled into a $563 billion global industry. The booming sector, defined as vacationing while enhancing one's physical, mental, or spiritual well-being, grew 14% in the last two years, nearly double that of overall tourism, reports the Global Wellness Institute. More than 690 million wellness-focused trips, ranging from bootcamps to meditation retreats, were taken in [2018]. (Raphael, 2017)

This trend has been particularly concentrated in countries with a preestablished tourism sector, with many of medical tourism and travel for treatment's popular travel destinations offering their own comprehensive wellness resorts catering to this demand; like in Spain, Costa Rica, and the Philippines, just to name a few. In that same vein, many of these resorts have begun offering traditional medical experiences in this resort setting, including noninvasive alternatives to cosmetic treatments, the implementation of novel body-scanning technology, blood and telomere analyses, and the introduction of "programmes for revitalisation, stress reduction, pain management" (Burrows, 2019). Some of these resorts even reportedly began harnessing the powers of technology sourced from and used by high-level government agencies and respected institutions like the National Aeronautics and Space Administration and Harvard University, like transcranial current stimulation and brain photobiomodulation.

These resorts have likewise sprung into prominence for their penchant to tackle one of society's most widespread, yet unaddressed underlying issues: sleep deprivation. According to the Sleep Foundation, 35.2% of all US adults do not get sufficient sleep on a daily basis—something that "has an estimated economic impact of over $411B each year in the United States alone" (Suni, 2020). Sleep problems are particularly prevalent among people already suffering from other mental health issues such as depression, anxiety, and bipolar disorder, oftentimes worsening their symptoms and overall mental state in the process. This ripples throughout society; sleep deprivation in physicians and medical technicians increases the risk of medical errors by 53% and has likewise been deemed responsible for thousands of automotive accidents a year (Bean, n.d.; Oxford University Press, 2021).

As such, it's only natural that many of these increasingly wellness-focused resorts have put a concentrated effort in tackling sleep deprivation problems throughout their amenities, with some offering high-end sleep recovery programs that use "polygraph to diagnose sleep patterns, nutritionist and general medical consultations, and psychotherapy, meditation and anti-stress sessions, among others" to help combat the problem (Burrows, 2019).

For some, wellness resorts offer the opportunity to receive life-altering treatments that have not yet been approved or greenlighted in their original countries of residence, much like stem cell therapy. Though banned in the United States aside from a short list of rare medical problems, stem cell therapy has shown plenty of promise as a tool in regenerative medicine, helping reduce signs of aging and other body-related stresses, and has become a natural inclusion in many of these wellness resort facilities abroad.

Much of these resorts' offerings are exactly what the modern medical tourism and travel for treatment industry hope to provide patients from around the world: preventative care that helps mitigate further issues down the road, tackling problems before they catapult into larger, health-altering complications.

That's not to say these experiences are accessible to everyone; luxury has a price, and as previously explored, Millennials have not had the simplest route toward financial success in the years following the 2008 economic crisis, even prior to the pandemic—an issue to be explored in further in depth later on. Nevertheless, Millennials' demands for the wellness tourism economy were heard loud and clear by international tourism as a whole in the years leading up to the pandemic, requisites that are now expected to continue their upward trajectory in prominence as the world transitions into the post-COVID-19 wellness economy.

8.2 WELLNESS RESORTS IN PANDEMIC AND POSTPANDEMIC TRENDS

Everything changed when the COVID-19 pandemic hit; travel opportunities became sparse, if not outright prohibited, health came to the forefront of the societal psyche, and the wellness tourism industry was required to adapt in turn. One distinct new trend has been the rise of immunity-boosting retreats, looking to improve attendees' day-to-day livelihoods and overall health with programs and amenities designed to specifically contribute toward the improvement of the immune system (Hill, 2021).

Many of these programs have been highly individualized, tapping into the luxury travel market's growing demand for personalization through curated treatments specific to each attendee. This has included on-demand nutritionists, purifying and cleansing treatments, and of course, clarity-centric offerings like yoga, meditation, and breath-work classes. As with many aspects of medical tourism, the amenities may vary depending on a destination's specific natural wonders, including access to local minder hot springs, or scenic hiking routes.

According to wellness travel specialist Dawn Oliver,

> There are lots of people in the world right now, maybe living by themselves, who have been home for so long that they're just starving for connection. Sure, they can jump on a Zoom call to try and feel connected, but that's very limited. People are looking for an intimate retreat experience, where they can not only have connections with one another but also that one-on-one connection with a teacher. (Jelski, 2021)

This has coincided with the new introduction of long-term COVID-19 recovery retreats, a burgeoning sector of the luxury tourism industry expected to make a continued impact on global tourism's postpandemic bounceback. Considering the virus's lingering effects, many of which are still to be discovered due to COVID-19's inherent novelty, those with change to spare and a penchant for wellness travel, particularly Millennials, are likely to be enticed in by this growing sector of the leisure and travel industry.

While in-depth research on the prevalence of long Covid is still in its early stages, a recent study by the U.K.'s Office of National Statistics found that 13.7% of its 20,000 respondents, all of whom had tested positive for COVID-19, still felt symptoms after three months. If this percentage is an indicator of the numbers elsewhere, more than 23 million people could be experiencing long Covid worldwide. (Schalkx, 2021)

Potential issues include "shortness of breath with exertion, dizziness, concentration difficulties, constipation, and intolerance to heat and cold," as well as kidney issues, organ damage, diabetes inflammation, and chronic fatigue. But one of the most prevalent issues in post-COVID-19 recovery has been brain fog, a cerebral feeling of fuzziness and sluggishness that makes it difficult to accomplish tasks or perform actions that were previously thought of as simple and straightforward. Despite this issue's clear commonality, there's been no medication or pill yet introduced to directly fix enduring brain fog. Enter clarity-focused wellness travel, a new sector of luxury travel looking to combat brain fog at its roots and restore travelers' cerebral capacities back to their former, previrus levels.

Rather than rely on medical marvels, most of these resorts' long-haul COVID-19 programs instead focus on the holistic route supplemented by state of the art technology, taking alternative approaches "that are either unproven in clinical settings or otherwise overlooked by Western medicine" (Schalkx, 2021). While attacking each individual's specific problems with a personalized schedule, common treatments include "bowel cleansing, energy therapy, lymphatic drainage treatments, and cryotherapy," the use of colonics, gut-health-focused diets, and anti-inflammatory interventions.

These programs certainly do not run cheap: recent figures show most of these intense post-COVID trips cost upward of $7000, often catapulting into the five-figure range. Interestingly enough, many of these programs are built on the back of these wellness resorts' preexisting arsenal of treatments, as some of COVID-19's symptoms—like brain fog—are common side effects of other ailments.

> Although long Covid is a novel phenomenon, the isolated symptoms are not. Treatments for brain fog and respiratory problems have been a mainstay in wellness brochures for decades; they run the gamut from Ayurvedic methodologies to science-backed technical approaches. This allowed wellness resorts to launch their Covid recovery programs relatively swiftly, without the need for new specialists or equipment. (Schalkx, 2021)

That being said, considering COVID-19 is anticipated to never truly go away and is eventually expected to be regarded as common to contract as modern Influenza (Stieg, 2021), these long-haul COVID-19 programs offered by luxury resorts do not seem to be going anywhere anytime soon, offering consumers the chance to experience exactly what they are demanding: clarity.

8.3 THE IMPLICATIONS OF CLARITY AS A LUXURY COMMODITY

Though hospitals and medical facilities have done their best to adapt to the needs of COVID-19 patients in this time of crisis, luxury resorts and their wellness programs provide an undeniably more comprehensive treatment to their attendees, considering their capabilities to address issues like diet, nutrition, exercise, and mental wellness—things hospital's COVID-19 wings unfortunately are not equipped to handle. But with the high-rolling price tag of these wellness retreats inherently prohibiting lower income folks from reaping the benefits of their programs, the question must be asked: has post-COVID clarity become an overpriced commodity completely unattainable by the working class?

To some, the answer seems to be an overwhelming "yes." The wealthy are already disproportionately favored to have a better outcome after contracting COVID-19, able to afford treatments at private institutions that those on public insurance (or are uninsured) cannot justify, likewise possessing the financial cushion to take time off of work to attend one of these multiday programs or seek treatment that lower income individuals simply cannot afford. The same extends into nutrition; organic and natural products cost more on average, while the majority of attainable cheaper options—many of which are unfortunately looked down upon as "lower class"—do not have the same health benefits, or worse, actually negatively contribute to a person's health status. Meanwhile, it's been well documented how many essential workers tending to these hospitalized individuals have passed away from their own COVID-19 exposure; meanwhile, the rich are able to recover in the lap of luxury.

This is particularly intriguing, as it's been previously explored how Millennials and young adults—the very people demanding these alternative treatment opportunities—have been disproportionately affected by the COVID-19 pandemic's economic impacts. With the large majority of this generation unable to afford these programs and luxury travel in general, the prevalent worldwide wealth gap, once thought of as a battle of young-versus-old, is becoming an increasingly obvious issue within consumer's very own generations.

So with these alternative long-haul COVID-19 programs, and postpandemic luxury wellness travel in general, rising as opportunities only attainable for the wealthy, it's becoming clear that alternative wellness treatments need to be reshaped and repackaged to become accessible to the general public.

"If we can get to a position where we find the elements of long COVID retreats which are the actual benefits—not just rich people on a holiday—then we can actually figure out ways to bring that to the masses," summed up COVID-19 researcher and the British Medical Association's Dr. David Strain (Pitcher, 2021).

KEYWORDS

- **medical tourism**
- **travel for treatment**
- **mental health**
- **COVID-19**
- **wellness**
- **clarity**

REFERENCES

American Psychological Association. *Low Socioeconomic Status Is a Risk Factor for Mental Illness*, March 6, 2005. https://www.apa.org/news/press/releases/2005/03/low-ses

Bean, M. *Moderate Sleep Loss Boosts Risk of Medical Errors 53%, Study Finds*, n.d. https://www.beckershospitalreview.com/patient-safety-outcomes/moderate-sleep-loss-boosts-risk-of-medical-errors-53-study-finds.html

Burrows, V. *How Millennials Are Embracing Holistic Luxury Getaways*, September 21, 2019. https://www.scmp.com/magazines/style/well-being/article/3028004/rise-wellness-retreats-how-millennials-are-embracing

Dimon, A. *How Millennials Are Disrupting the Wellness Travel Market*. Travel Market Report: The Voice of the Travel Advisor, 2018. https://www.travelmarketreport.com/articles/How-Millennials-Are-Disrupting-the-Wellness-Travel-Market

Hill, L. J. *The Rise in Immunity Boosting Wellness Retreats for 2021*. Forbes, January 4, 2021. https://www.forbes.com/sites/laurenjadehill/2021/01/04/the-rise-in-immunity-boosting-wellness-retreats-for-2021/?sh=8e1410e7b5d7

Jelski, C. *Resurgent Retreats*; Travel Weekly, March 3, 2021. https://www.travelweekly.com/Travel-News/Hotel-News/Resurgent-retreats

Kaiser Family Foundation. *The Implications of COVID-19 for Mental Health and Substance Use*, April 14, 2021. https://www.kff.org/coronavirus-COVID-19/issue-brief/the-implications-of-COVID-19-for-mental-health-and-substance-use/

Oxford University Press. *Sleep Deprived People More Likely to Have Car Crashes*. Science Daily, October 24, 2021. https://www.sciencedaily.com/releases/2018/09/180918082041.htm

Pitcher, L. *Post-COVID Wellness Retreats: Disaster Capitalism or a Helpful Resource?* Dazed, November 4, 2021. https://www.dazeddigital.com/beauty/body/article/54680/1/post-COVID-19-wellness-retreats-disaster-capitalism-or-a-helpful-resource

Raphael, R. *Travelers Are Abandoning Spas to Join Fitfluencer Retreats*; Fast Company, December 9, 2017. https://www.fastcompany.com/40504343/travelers-are-abandoning-spas-to-join-fitfluencer-retreats

Schalkx, C. Wellness Resorts Eye Long COVID Treatments as New Moneymaker. *Bloomberg*, July 13, 2021. https://www.bloomberg.com/news/articles/2021-07-13/long-covid-treatments-crop-up-at-luxury-wellness-resorts

Stieg, C. The Pandemic Could End in 2022—here's What 'Normal' Life Might Look Like Soon, According to Medical Experts. *CNBC*, December 15, 2021. https://www.cnbc.com/2021/12/15/what-living-with-endemic-covid-will-look-like-in-2022-and-future.html

Suni, E. *Sleep Statistics*; Sleep Foundation, December 11, 2020. https://www.sleepfoundation.org/how-sleep-works/sleep-facts-statistics

US Bureau of Labor Statistics. *Effects of the Coronavirus COVID-19 Pandemic (CPS)*, May 18, 2021. https://www.bls.gov/cps/effects-of-the-coronavirus-COVID-19-pandemic.html

United Nations. *World Youth Report: Youth Social Entrepreneurship and the 2030 Agenda*, 2020. https://www.un.org/development/desa/youth/wp-content/uploads/sites/21/2020/07/2020-World-Youth-Report-FULL-FINAL.pdf

World Economic Forum. *The Global Risks Report 2021* (16), 2021. https://www3.weforum.org/docs/WEF_The_Global_Risks_Report_2021.pdf

CHAPTER 9

ENDING YOUR LIFE IN ANOTHER COUNTRY

ABSTRACT

The debate on physician-assisted suicide (PAS) and its place in healthcare is filled with contention and has yet to reach a clear consensus. This chapter explains the difference between PAS and euthanasia, provides insight into the widely varying policies on the matter throughout international countries and US states and an exploration of arguments for and criticisms against PAS and some of the topic's most notable cases, as well as information on its present role in the medical tourism industry at large.

While medical tourism and travel for treatment have been hailed across the global community as live-saving opportunities for medical patients, this global sector of the health industry can be utilized for the opposite purpose as well: physician-assisted suicide, also known as PAS. Though the laws vary from country to country, where PAS is in fact legal, it is usually accomplished by a doctor writing a prescription for a lethal dose of pharmaceuticals, rather than a person ending their lives themselves.

This stands in contrast to euthanasia—a term many critics of PAS incorrectly use interchangably with POS—where a person's life is ended to stop their suffering, as euthanasia has historically be performed both voluntarily and without volition. This is usually referred to as voluntary euthanasia in cases where consent has been granted, and as nonvoluntary euthanasia when consent is missing and the euthanasia is performed unknowingly to the patient.

Even the name "assisted suicide" has come under fire for the use of the word "suicide," with supporters of the PAS cause preferring to reference the term as "assisted dying" or "medical aid in dying." This provides a

Medical, Dental, and Wellness Tourism: A Postpandemic Perspective. Mary Schreiber Swenson, PhD, and Amit Bansal (Authors).

distinction between terminally ill persons being allowed to choice whether they carry on with living (assisted dying) and people of any health status to end their lives (assisted suicide) (Moss, 2013). Given the contentious nature of verbiage surrounding it, we will refer to the act as "physician-assisted suicide" moving forward for the sake of simplicity and understanding.

Despite numerous vocal advocates for the cause, in many countries, PAS is not only frowned upon, but altogether banned and considered a crime (Patients Rights Council, n.d.). Doctors that aid patients with ending their lives, no matter the terminal status of a chronic disease or other painful health issues, can end up on trial for manslaughter or other similar charges in spite of the fact that their services were contracted and agreed upon by the patient.

Still, a number of countries still continue to offer PAS within their borders, including Spain, Germany, Switzerland, Austria, Belgium, the Netherlands, Canada under specific cases, and a few select US states. While ending life with dignity may not be an option in a person's original country of residence, medical tourism and travel for treatment can open the door for terminal patients and persons of all backgrounds to seize the fate of their lives back into their own hands.

9.1 THE CONTROVERSIES AND CRITICISMS OF PAS

To understand the full scope of PAS and its role in international society, it's vital to understand the two sides of the PAS debate—"right to life" and "right to die"—as arguments may vary based on different religious, ethical, and regional standards.

In the United States, the American Medical Association's Code of Ethics plainly states that PAS is prohibited for doctors falling underneath its jurisdiction, saying that "physician-assisted suicide is fundamentally incompatible with the physician's role as healer, would be difficult or impossible to control, and would pose serious societal risks" (Riddick, 2003). This has been backed up by the court of law, as the US Supreme Court ruled in 1997's *Vacco v. Quill* that the state-by-state creation of laws banning PAS is not in fact unconstitutional (Lagay, 2003), and later that same year in *Washington v. Glucksberg* that the right to die is not granted by the Due Process Clause (Dugdale et al., 2019).

Still, while states are able to make their individual decisions about whether or not they will allow PAS, the Supreme Court did not go so far as to ban the practice on a national level, preferring to keep the issue one of state's rights.

As a result, there are ten states and one district across the United States that have legalized PAS: California, the District of Columbia, Colorado, Hawaii, Montana, Maine, New Jersey, New Mexico, Oregon, Vermont, Washington, and, due to a court ruling, Montana, all with their own unique provisions to these laws (Procon.org, 2019).

Detractors of PAS also point to the Hippocratic Oath, the classic ethical oath taken by physicians that dates back to the third century BCE, in which physicians swear "I will not give a lethal drug to anyone if I am asked, nor will I advise such a plan" (Hippocrates). However, given the Hippocratic Oath's continual evolution and the changes made to it across the course of the millennia since its foundation, and likewise the lack of requirement for physicians to follow said oath in medical settings, using the Hippocratic Oath as an argument against PAS is a debate built on shaky ground.

Attitudes against PAS can also be traced back to the international climate post-World War II, as the atrocities of eugenics and euthanasia committed by the Nazi German regime led the World Medical Association to draft a new version of the Hippocratic Oath itself in 1948 through the Declaration of Geneva, which was later revised in 1968, 1983, 1994, 2005, 2006, and 2017 (World Medical Association, 2017). This includes additions like "I will maintain the utmost respect for human life," which many use as part of their reasoning against the morality and legality of PAS.

This is similarly reflected in the International Code of Medical Ethics, which is based on the Declaration of Geneva and was implemented by the World Medical Association in 1949, which plainly states that "a physician shall always bear in mind the obligation to respect human life" (World Medical Association, 2010). While initially established in the 1940s, this attitude toward PAS endured throughout the 20th century, as 1992's 44th World Medical Assembly said that "physician-assisted suicide, like voluntary euthanasia, is unethical and must be condemned by the medical profession" in its Statement of Marbella (World Medical Association, 2014).

Many physicians who have aided in illegal assisted suicide have been criminal prosecuted for their practices in countries where it is not legal, even if their choices were ethically motivated, and can be charged with manslaughter or homicide in their role in PAS.

In the 1980s, the pathologist Jacob "Jack" Kevorkian began advertising in Detroit area newspapers as a death counselor. He had studied the technique of Dutch physicians in the Netherlands, and created his own device with which patients could self-administer lethal medications. His first patient ended her life in 1990 while lying on a bed inside Kevorkian's Volkswagen

> van. He went on to assist with some 130 deaths by suicide over the next eight years. In 1999, after Kevorkian publicly distributed a video of himself directly euthanizing a patient, he was convicted of second-degree murder and sent to prison. (Dugdale et al., 2019)

Religion has also played a significant role in worldwide perceptions of the PAS cause, as certain doctrines and religious texts have explicit passages detailing the importance of preserving life until it ends naturally. This is particularly evident in religions that trend toward fundamentalism or literal interpretation of aforementioned texts, like Roman Catholicism and the Church of Jesus Christ of Latter-day Saints (n.d.).

Euthanasia and PAS are vocally fought against by the LDS, though the church does make some exceptions, saying that "when dying becomes inevitable, death should be looked upon as a blessing and a purposeful part of an eternal existence. Members should not feel obligated to extend mortal life by means that are unreasonable. These judgments are best made by family members after receiving wise and competent medical advice and seeking divine guidance through fasting and prayer" (Church of Jesus Christ of Latter-Day Saints, n.d.).

Catholicism has been more staunch in its anti-PAS stance, believing that suffering experienced during mortal life does not give humanity the choice to decide when life ends over the will of the church's god, holding to scripture that says "God is the creator and author of all life." This also entails the belief that PAS also denies friends and families further time to spend with the patient whose life ended and causes them to experience unnecessary grief (Donovan, 1997). The Roman Catholic Church's religious figurehead, Pope Francis, has publicly endorsed the anti-PAS belief system, citing the biblical Ten Commandments which dictate "thou shalt not kill" and claiming that assisted suicide directly contradicts the teachings of the Roman Catholic god. With more than 1.3 billion Roman Catholics around the world—nearly one-fifth of the global population—the church's beliefs against PAS are palpable across the corners of the map (WorldData.info, n.d.).

Judaism similarly holds up the preservation of human life as one of its core tenants, and most Jewish religious authorities and Rabbis believe that assisted suicide is an inappropriate course of action regardless of a patient's terminal status. However, there have been notable instances of assisted suicide throughout Judaism's older texts, potentially leaving acceptance of the process open to interpretation depending on a person's level of orthodoxical beliefs.

Beyond religious and medical doctrine, there remains a large group of chronically ill patients who remain advocates against PAS, citing the "slippery slope" effect as reasoning. This states that accepting PAS as a norm could open the door for involuntary euthanasia down the line, though critics of the "slippery slope" mentality claim these concerns can be combated through legal protections and safeguards, as demonstrated in Belgium and the Netherlands, where PAS for specific terminal illnesses is legal.

> In Belgium, the 'patient [must be] in a medically futile condition of constant and unbearable physical or mental suffering that cannot be alleviated, resulting from a serious and incurable disorder caused by illness or accident'. Like the Netherlands, there is no requirement that the patient be suffering from a terminal illness, although additional procedural requirements are imposed if the patient is 'clearly not expected to die in the near future'. Again there must be 'no reasonable alternative' to euthanasia. However, euthanasia is permissible only if the disorder is incurable, so a patient's reasonable refusal of potentially curative treatment will prevent access to euthanasia. The reasonable refusal of a palliative treatment possibility will not have this effect. (Lewis and Black, 2012)

The Netherlands provides physician-assisted euthanasia for infants under the age of one through 2004's Groningen Protocol as mandated by the Dutch Society for Pediatrics, which lays out the acceptance for this process under four important criteria, all of which must be fulfilled: the presence of hopeless of unbearable suffering, parental consent to terminate life, completed medical consultations, and careful execution of the termination (Vizcarrondo, 2014). Any infant euthanasia in the Netherlands must be reported by the practicing physician to the local medical examiner and later reported and reviewed by a review committee and local district attorney. While the majority of cases in the Netherlands have been prompted by an infants' incurable and life-altering disease, like spina bifida and hydrocephalus, many decorators of PAS—particularly those stemming from religious ideologies—have characterized the Groningen Protocol as infantacide, though opinions vary throughout the medical community at large (Verhagen et al., 2005). Though a 1999 study argued that the Netherlands' PAS and euthanasia practices were underregulated and not well controlled, a follow-up review of the Netherlands' two-decade-long practice of these procedures by the Netherlands Department of Public Health in 2009 found there to be no evidence of the slippery-slope effect having taken place (Rietjens et al., 2009).

9.2 ARGUMENTS IN FAVOR OF PAS

In spite of its many detractors, there remains a consistent base of supporters of PAS across the global medical community and general population. Many people believe that it is up to an individual's personal autonomy to decide whether they want to carry on with life—also known as the "right to die" argument—and in line with principles of self-sovereignty.

Perhaps the most impactful voices on PAS are terminal patients themselves, whose sharing of their personal anecdotes, medical histories, and chronic pain has led to greater understanding on *why* terminally ill individuals can find relief in PAS and led to some legislational changes in favor of aid in dying (AID) it turn.

> In 2014, a young Californian named Brittany Maynard was diagnosed with an astrocytoma and became a spokesperson for the legalization of AID. She was a newlywed facing terminal illness, and her story quickly captured the public imagination. Her well-publicized death by lethal ingestion in Oregon in 2014 influenced her home state of California to legalize AID in 2015. This was subsequently followed by Colorado in 2016, the District of Columbia in 2017, Hawai'i in 2018, and New Jersey and Maine in 2019. (Dugdale et al., 2019)

PAS has likewise been advocated for by a multitude of bioethicists, who study the greater ethical, social, and legal impacts of the medicine and the medical industry at large. According to bioethicist Tom Beauchamp and James Childress, the four principles behind bioethics are autonomy, beneficence, nonmaleficence, and justice, all of which are applicable to the concepts of PAS and AID, depending on a case's unique context. However, autonomy is the most common principle used by this school of thought as an argument in favor of PAS, believing that "patient autonomy serves as the justification for informed consent; only after a thorough explanation of risks and benefits can the patient have the agency to make a decision about treatments or participation in medical research. This logic, it is argued, naturally extends to AID; patients accustomed to making their own health care decisions throughout life should also be permitted to control the circumstances of their deaths" (Dugdale et al., 2019).

The pro-PAS approach goes on to address the Hippocratic Oath's seemingly anti-PAS provisions, claiming that helping people relieve their pain and suffering through lethal injection or other forms of aid in dying is both humane and an act of compassion. This argument was prominently used by Maynard in her struggles to receive AID in California, as well by other terminally ill individuals fighting for their right to die.

Advocates likewise point to PAS as a medically safe alternative to self-inflicted suicide, as doctors can ensure that the procedure is complete until termination. According to research conducted by Harvard University, 9 out of 10 suicide attempts fail. While not all of those attempts are made by terminally ill patients looking to end their lives with dignity, the statistics overwhelmingly show that self-inflicted suicide fails the large majority of the time (Harvard University, 2012). Likewise, lingering effects from failed suicide attempts can extend or amplify and individual's pain and suffering. As such, PAS can eliminate this margin from error for people who qualify, providing them with a securely administered and carefully monitored way to die performed by a medical professional that's protected by legal safeguards.

Many point to Oregon, one of the 11 US states and districts that has allowed PAS, as a positive example of what introducing AID can do for an area's terminally ill residents. The state's 1997 Death With Dignity Act, as voted into effect by a majority of the state's constituents, legalized PAS under certain restrictions and allows competent terminally ill adults with less than six months to live to receive a prescription lethal dose of medication from their physician, though physicians and pharmacists are allowed to refuse PAS participation based on their own moral and religious beliefs. PAS requests must be backed up by two witnesses, including one with no relation or financial ties to the patient, which are then reviewed by a medical professional to confirm diagnosis and ascertain a PAS candidate does not have any mental conditions that could influence or sway their decisions (Oregon State Government, n.d.).

Though there have been multiple attempts to repeal the law, the Death with Dignity Act's continued legality and apparent efficacy has been used by PAS proponents across the world. Between 1997 and 2021, 1905 patients followed through on the Death with Dignity Act's provisions through ingesting the lethal medications prescribed to them, with the majority of these individuals having been diagnosed with terminal cancer (Oregon Public Health Division, Center for Health Statistics, 2019). A 2007 independent study and analysis of Oregon and the Netherlands' PAS laws showed that vulnerable populations are not unduly subjected to AID provisions, refuting opponents' slippery-slope claims once more.

> Rates of assisted dying in Oregon and in the Netherlands showed no evidence of heightened risk for the elderly, women, the uninsured (inapplicable in

> the Netherlands, where all are insured), people with low educational status, the poor, the physically disabled or chronically ill, minors, people with psychiatric illnesses including depression, or racial or ethnic minorities, compared with background populations. The only group with a heightened risk was people with AIDS. While extralegal cases were not the focus of this study, none have been uncovered in Oregon; among extralegal cases in the Netherlands, there was no evidence of higher rates in vulnerable groups….
>
> ….Where assisted dying is already legal, there is no current evidence for the claim that legalised PAS or euthanasia will have disproportionate impact on patients in vulnerable groups. Those who received physician-assisted dying in the jurisdictions studied appeared to enjoy comparative social, economic, educational, professional and other privileges. (Battin et al., 2007)

Demographic information from Oregon's Death with Dignity Act annual data summary backs up these claims; 81% of these individuals were over the age of 65, and 97% of them were white, refuting the idea that PAS can be a slippery-slope into eugenics or involuntary euthanasia. Additionally, Oregon's analysis revealed that for most of the patients who followed through on the Death with Dignity Act echo the same sentiments as many pro-AID supporters, as "the three most frequently reported end-of-life concerns were decreasing ability to participate in activities that made life enjoyable (94%), loss of autonomy (93%) and loss of dignity (72%)" (Oregon Public Health Division, Center for Health Statistics, 2019). Advances in Oregon's PAS framework helped open the door for other US states to similarly adopt their own variation of PAS legality into legislation.

Certain religions have adopted the "right to die" approach into their doctrines, particularly seen in Hinduism and Jainism, which are primarily practiced in India. Though outside the realm of PAS, these religions believe in nonviolent suicide through the form of fasting, known as Prayopavesa and Santhara.

Taking a more Western approach, the religion of Unitarian Universalism announced its advocacy for the right to die in its 1988 General Resolution, stating that "Unitarian Universalists advocate the right to self-determination in dying, and the release from civil or criminal penalties of those who, under proper safeguards, act to honor the right of terminally ill patients to select the time of their own deaths" (Unitarian Universalist Association, 2011).

9.3 MEDICAL TOURISM AND TRAVEL FOR TREATMENT'S ROLE IN PAS

Considering the legality of PAS and aid in dying varies widely by country to country—and even state by state, in the case of the United States—many terminally ill patients have been looking abroad through the lens of medical tourism and travel for treatment to find an adequate destination to follow through on their self-determined right to die. While many attempts to distance the medical tourism industry away from PAS, claiming that the process is a far cry from the cost-saving attractions of traditional medical tourism (Harkins, n.d.), the core principle of international PAS is the same: traveling to another location to receive care that's unattainable in their country of origin. An undeniably complex issue, it is become increasingly common for medical tourism agencies and thought-leaders to field requests from patients looking to end their lives in dignity, and as such it is important to explore what role medical tourism and travel for treatment play in these difficult right-to-die situations.

"Throughout my career, I have had only a small number of people come to me to find a solution in ending their very painful life," says Dr. Mary Swenson about her own personal experience with PAS throughout her time as CEO of MyMedChoices. "The pain stems from a very rare form of cancer or injury that puts the patient in so much pain that they are unable to live their daily lives. If someone would ever ask to end their lives due to an emotional issue, I would turn them down as ethics plays a major role in this request. The term is often referred to as 'suicide tourism' and people many travel to Zurich, Switzerland where it is legal. When planning this journey, it becomes a very personal journey for the patient as family is involved, and it is very emotional for not only the medical tourism facilitator but the family and patient." continued Swenson.

"In my career, I have had around 11 people approach me about this request and of the 11, only one gentleman went through with the request," continues Swenson. "He was 98 and was diagnosed with pancreatic cancer and was in great pain. He lived a very full life with his family and he requested that they not see him suffer. It took over two years to plan the journey and he traveled with frail weakness. Assets were divided with no family feuding and the grandest of parties was thrown in his honor."

"Whether your moral compass allows you to accept this form of tourism or not, it is an option for those who choose," concluded Swenson.

Switzerland has become one of the main destinations for suicide tourism across the 21st century as PAS has been allowed there since 1942, with a documented 611 people traveling to the country between 2008 and 2012 to follow through on their self-made decision to end their lives, with many of these travelers hailing from the United Kingdom, Germany, and other nearby European countries (Gauthier et al., 2015). According to a study conducted by the Institute of Legal Medicine at the University of Zurich, "in the UK, at least, 'going to Switzerland' has become a euphemism for (assisted suicide). Six right-to-die organizations assist in approximately 600 cases of suicide per year; some 150–200 of which are suicide tourists." The study found that the majority of these tourists were suffering from some form of neurological disease, followed by cancer, heart disease, and rheumatic diseases.

While previously explored AID infrastructure in locations like Oregon require patients to be diagnosed as terminal and close to death, the increase in suicide tourists looking to end their lives over neurological diseases in destinations like Switzerland paints a picture of the pain and suffering the deterioration of the brain can bring upon a person. Diseases like Parkinson's, Alzheimer's, and multiple sclerosis can affect people's autonomy, as these conditions often come with debilitating symptoms like partial to complete memory loss, decreased limb function, and mobility issues, which can affect their overall quality of life (Pan American Health Organization, n.d.). As these diseases continue to chip away at people's self-sufficiency and autonomy as they continue to develop in the brain, it's not difficult to see why patients with these chronic neurological illnesses wish to end their lives in destinations where PAS is legal while their fate is still in their own hands, or before they become a shell of the person they once were, especially as the aforementioned diseases are still without a cure. As a result, medical tourism or suicide tourism is often the only choice for those with neurological diseases to die with dignity.

> Seven themes emerged from the data analysis, including health and functioning; feelings regarding survivorship and existence; interacting with the health sector; attitudes regarding death and dying; suicide; choosing death; and choosing suicide tourism. A significant portion of the participants had experienced suicidal thoughts and had even previously attempted suicide, some more than once. Most of them referred to chronic illnesses, functional disability, and social isolation. They understand suffering within the subjective dimension, namely only by the person who is actually subjected to the disease, ailments, and disability. Participants regarded aid-in-dying in Switzerland as positive thanks to its guaranteed outcome: "beautiful

> death," compared to "disadvantaged dying" which places a burden on the participants' loved ones throughout the prolonged dying. Most of them do not necessarily want to have their loved ones beside them when they die, and they see no significant meaning in dying in a foreign country to which they have no emotional or civil attachment. (Sperling, 2022)

Though a clearly contentious subject, medical tourism and travel for treatment can provide one of the only avenues for the right to die to people whose lives are inextricably affected by disease. Medical tourism can help set these travelers up with trained physicians in a fitting destination to follow through on this decision, ensuring the right to die is taken safely and performed effectively, far removed from the statistical failures of self-inflicted suicide previously explored. The suicide tourism debate is only anticipated to become increasingly important in the future as the rate of chronic disease continues to grow among the globe's aging population (Worley, 2007), as can be seen by the recent legalization of the practice across a number of international countries.

9.4 LEGAL DESTINATIONS FOR PAS

Australia: While not federally legalized across the country, the Australian government has made the subject of PAS a matter of states' choice and is currently legal across all of Australia's states through their own region-specific legislation while remaining prohibited in the Australian territories. States like New South Wales legalized the practice as recently as May 2022 (Pan American Health Organization, 2019).

Austria: Following a ruling by the Austrian Constitutional Court that outlawing PAS is unconstitutional, Austrian Parliament legalized the practice in 2021 for patients with terminal or debilitating conditions. Each Austrian case will be examined by two doctors and strictly regulated, and the provisions exclude minors and those with mental health conditions (BBC News, 2022).

Belgium: Belgium has one of the world's most lax voluntary euthanasia laws through 2002's Euthanasia Act, though it refrains from explicitly naming the practice as PAS. This extends right-to-die opportunities to minors and those with mental illnesses and is not exclusively reserved for terminal cases as in other parts of the world, and AID has been given popular support from

Belgian residents themselves (Patients Rights Council, PBS News Hour, 2015).

Canada: Canada's Supreme Court declared PAS legal nationwide in 2015's *Carter vs. Canada*. In March 2021, Canada introduced new provisions to the medical assistance in dying law to broaden PAS safeguards and eligibility requirements, which will introduce mental illness as a viable reason under certain criteria to receive AID in March 2023 (Harvard University, Health Canada, 2018).

Colombia: Voluntary euthanasia has been legal in Colombia since 1997, and recently began allowing people with non-terminal illnesses to engage in AID, as evidenced by the 2022 case of Victor Escobar, who elected into his right to die due to effects of his end-stage chronic obstructive pulmonary disease. "We reached the goal for patients like me, who aren't terminal but degenerative, to win this battle, a battle that opens the doors for the other patients who come after me and who right now want a dignified death," said Escobar (Reuters, 2022). The Colombian Supreme Court has safeguarded the right to AID with a decision that "spelled out the rights of a terminally ill person to engage in voluntary euthanasia."

Denmark: Though assisted suicide is illegal across Denmark, there are legal provisions for passive euthanasia—also known as the refusal to accept treatment that can extend a life. However, Danish citizens are overwhelmingly in support of PAS, with 71% of the population in favor of legalizing the practice in Denmark (Nielsen and Andersen, 2014).

Germany: PAS became formally legalized in Germany following a court ruling in 2020 that overturned "commercial promotion of assisted suicide" (BBC News, 2020).

Italy: Assisted suicide was legalized by the Italian Constitutional Court in September 2019, though the government has yet to formalize regulations around the practice. Italy's first legal assisted suicide was performed in June 2022 (Palmer, 2022).

Jersey: Jersey became the only place in the British Isles to allow assisted dying in a decision made in November 2021 on the back of widespread public support, allowing an adult with the proven mental capacity to choose the right to die as long as they have "been diagnosed with a terminal illness and has a life expectancy of 6 months or less, or has an incurable physical condition that causes enduring and unbearable suffering" (States of Jersey, n.d.).

Luxembourg: The European country formally legalized assisted suicide in 2008 under the Law on the Right to Die with Dignity. The law provided regulatory bodies like Luxembourg's National Commission of Control and Evaluation to safeguard the practice, and offers provisions for minors between the ages of 16 and 18 so long as they receive parental authorization and meet the law's other requisites (Library of Congress, n.d.).

Netherlands: As previously explored, the Netherlands has become one of the most influential countries on the subject of PAS after becoming the first European country to legalize the practice in 2001 under the "Termination of Life on Request and Assisted Suicide (Review Procedures) Act," expanding the country's previously legalized voluntary euthanasia practices. The Netherlands allows minors above the age of twelve to receive PAS, as long as they meet the same diagnoses of suffering "unbearably without hope" adults need to receive from physicians, and likewise includes provisions for suffering infants under the Groningen Protocol (Alliance VITA, 2017).

New Zealand: New Zealand's End of Life Choice Act 2019 went into effect in November 2021 after a citizen referendum ratified the policy, which allows terminally ill people in the country to receive AID. The requirements to receive PAS are slightly stricter in New Zealand than some other countries and requires patients to meet all of six specific criteria—including that they must suffer from a terminal illness that is likely to end their life within six months, they must be in an advanced state of irreversible decline in physical capability, and they must experience unbearable suffering that cannot be relieved in a manner that they consider tolerable—before meeting approval (Ministry of Health New Zealand, n.d.).

Switzerland: One of the only countries that accepts nonresident international travelers for AID cases, Swiss law allows people of all backgrounds and medical conditions to receive assisted suicide as long as they have the capacity to make autonomous decisions, and likewise allows for non-PAS. This has made Switzerland one of the prominent choices for suicide tourism due to its flexible laws and its inclusion of foreigners (Andorno, 2013).

The United States: Ten states and one district have currently legalized or decriminalized assisted suicide on state-specific legislation and judgments, including California, Colorado, Hawai'i, Oregon, New Jersey, New Mexico, Maine, Montana, Vermont, Washington, and the District of Columbia. Most encompass the same basic requirements, which state that a patient needs to be a legal resident of its respective state, a legal adult, terminally ill, and of sound mind at the time of their decision. In states where the practice is not yet legalized, the punishments and penalties for a physician engaging in PAS vary from state to state, such as in Florida, where "every person deliberately assisting another in the commission of self-murder shall be guilty of manslaughter, a felony of the second degree" (Patients Rights Council, n.d.). Arizona, Connecticut, Indiana, New York, and Virginia are all considering adopting their own PAS legislation as of 2022.

California: First going into effect in June 2016, the California End of Life Option Act lets terminally ill Californians obtain a lethal dose of medication from their physician for the purpose of ending their own lives. Terminally ill patients must have been diagnosed with only six months or less to live, be over the age of 18, and have the physical and mental capacity to both make the decision and ingest the drug themselves (UCLA Health, n.d.).

Colorado: Colorado's Colorado End-of-Life Options Act was approved by Coloradoan voters in 2016, allowing terminally ill individuals with less than six months to live to obtain a prescription for a lethal dose medication from their physician for self-administration. Colorado's statute includes provisions that create criminal penalties for anyone who tampers with AID medication or coerces an individual into requesting it (Colorado Department of Public Health & Environment, n.d.).

Hawai'i: First passed in 2018, Hawai'i is now exploring expanding its Our Care, Our Choice Act with additional legislation. The current iteration of the Our Care, Our Choice Act has similar provisions to California and Colorado's, featuring requisites like being over the age 18, a terminal illness diagnosis with less than six months to live, and the mental capacity to make the choice for themselves, alongside being a resident of Hawai'i.

Oregon: Oregon's 1997 Death with Dignity Act is considered a groundbreaking piece of legislation for right-to-die advocates in the United States, and marks the first time a US state legalized AID. Recent updates to the law include Oregon Governor Kate Brown's 2019 decision to waive the 15-day waiting period to receive medications for patients who are imminently dying (Death with Dignity, n.d.).

New Jersey: New Jersey's Medical Aid in Dying for the Terminally Ill Act was first passed in 2019, allowing legal adults that are New Jersey residents to receive AID medication from their physician if they are mentally fit and terminally ill. New Jersey right-to-die patients will need to go through a number of hoops to receive their medication, including potential psychiatric evaluations, the need for multiple witnesses, and the case's confirmation by multiple doctors (Gillespie, n.d.-a).

New Mexico: The End-of-Life Options Act (HB47) was signed into New Mexican law in 2021, letting terminally ill New Mexican residents obtain medical aid in dying if they meet a number of strict criteria. This includes being of fit mental capacity to make a decision as well as a life expectancy of less than six months left. Under New Mexico law, a physician who helps a patient end their life may face criminal charges (Gillespie, n.d.-b).

Maine: Maine originally legalized PAS in June, 2019, becoming the eighth US state to do so. "The Maine Death with Dignity Act allows competent, terminally ill state residents who are within 6 (six) months of death to legally obtain oral prescription medication they can voluntarily take without assistance to peacefully end their life" (Maine Death with Dignity Organization, n.d.).

Montana: Montana is a unique state in the landscape of US death with dignity, as its provisions do not come from the introduction of legislation but from the decision of the Montana Supreme Court in 2009's *Baxter v. Montana*. Though this did not formally legalize PAS, it created precedent for physicians who provide AID to their patients to be protected from state prosecution. This could potentially be overruled by legislation, though the decision remains in place as of July 2022. Montana also has the Montana Rights of the Terminally Ill Act, which allows terminally ill patients the right to refuse treatment or medications at the end of life (FindLaw, n.d.).

Vermont: According to Vermont's Patient Choice and Control at End of Life Act (Act 39), Vermont law "provides eligible Vermont residents with terminal diseases the option to be prescribed a dose of medication that, if taken, will hasten the end of their life. This option requires the participation of a Vermont-licensed physician." Similar to other US states, patients need to be in a sufficient mental capacity to make the choice to end their lives, be capable of taking the medication themselves, and have been diagnosed with less than six months left to live (Vermont Department of Health, n.d.).

Washington: The second state to legalize PAS after Oregon, Washington passed its Death with Dignity Act in 2008 approximately eleven years after its neighbor's ground-breaking legislation with a majority vote from the Washington populus. This includes provisions like the need to be a Washington state resident, over the age of 18, terminally ill with less than six months to live, and the mental competence to make choices for themselves (Washington Government, 2009).

District of Columbia: Introduced into law in 2016, the District of Columbia's "Death with Dignity Act provides for District of Columbia residents, qualified with a terminal disease, to die in a humane and peaceful manner through the voluntary use of prescribed medications." As in other US jurisdictions, this act requires patients to be over the age of 18, a legal DC resident, of sound mind, and with a prognosis of less than six months to live (District of Columbia Department of Health, n.d.).

KEYWORDS

- **medical tourism**
- **travel for treatment**
- **euthanasia**
- **physician-assisted suicide**
- **assisted dying**
- **voluntary euthanasia**

REFERENCES

Alliance VITA. *Euthanasia in the Netherlands*; Alliance VITA, November 24, 2017. https://www.alliancevita.org/en/2017/11/euthanasia-in-the-netherlands

Andorno, R. Nonphysician-Assisted Suicide in Switzerland. *Cambridge Q. Healthcare Ethics* **2013,** *22* (3), 246–253. https://doi.org/10.1017/s0963180113000054

Battin, M. P.; van der Heide, A.; Ganzini, L.; van der Wal, G.; Onwuteaka-Philipsen, B. D. Legal Physician-Assisted Dying in Oregon and the Netherlands: Evidence Concerning the Impact on Patients in "vulnerable" Groups. *J. Med. Ethics* **2007,** *33* (10), 591–597. https://doi.org/10.1136/jme.2007.022335

BBC News. Germany Overturns Ban on Assisted Suicide. *BBC News*, February 26, 2020. https://www.bbc.com/news/world-europe-51643306

BBC News. New Law Allowing Assisted Suicide Takes Effect in Austria. *BBC News*, January 1, 2022. https://www.bbc.com/news/world-europe-59847371

Church of Jesus Christ of Latter-Day Saints. *Euthanasia and Prolonging Life*; Church of Jesus Christ of Latter-Day Saints, n.d. https://newsroom.churchofjesuschrist.org/official-statement/euthanasia-and-prolonging-life

Colorado Department of Public Health & Environment. *Medical Aid in Dying*. Cdphe.colorado.gov; Colorado Department of Public Health & Environment, n.d. https://cdphe.colorado.gov/center-for-health-and-environmental-data/registries-and-vital-statistics/medical-aid-in-dying

Death with Dignity. *Oregon*; Death with Dignity, n.d. https://deathwithdignity.org/states/oregon/

District of Columbia Department of Health. *Death with Dignity Program Frequently Asked Questions*. District of Columbia Department of Health, n.d. Retrieved December 3, 2019. From https://dchealth.dc.gov/sites/default/files/dc/sites/doh/page_content/attachments/Death%20with%20Dignity%20-%20Frequently%20Asked%20Questions%20%28FAQ%29.03.20.18.pdf

Donovan, G. K. Decisions at the End of Life: Catholic Tradition. *Christ. Bioethics* **1997,** *3* (3), 188–203. https://doi.org/10.1093/cb/3.3.188

Dugdale, L. S.; Lerner, B. H.; Callahan, D. Pros and Cons of Physician Aid in Dying. *Yale J. Biol. Med.* **2019,** *92* (4), 747–750. https://www.ncbi.nlm.nih.gov/pmc/articles/PMC6913818/

FindLaw. *Montana Euthanasia Laws*. FindLaw, n.d. https://www.findlaw.com/state/montana-law/montana-euthanasia-laws.html

Gauthier, S.; Mausbach, J.; Reisch, T.; Bartsch, C. Suicide Tourism: A Pilot Study on the Swiss Phenomenon. *J. Med. Ethics* **2015,** *41* (8), 611–617. https://doi.org/10.1136/medethics-2014-102091

Gillespie, J. *New Jersey's Medical Aid in Dying for the Terminally Ill Act*. Nolo, n.d.-a. https://www.nolo.com/legal-encyclopedia/new-jersey-s-aid-dying-the-terminally-ill-act.html

Gillespie, J. *New Mexico's End-of-Life Options Act*. www.nolo.com; Nolo, n.d.-b. Retrieved July 4, 2022. From https://www.nolo.com/legal-encyclopedia/death-with-dignity-new-mexico.html

Harkins, J. *Assisted Suicide: Who Decides—When, Where and How?* Medical Tourism Magazine, n.d. Retrieved June 29, 2022. From https://www.magazine.medicaltourism.com/article/assisted-suicide-decides

Harvard University. *Attempters' Long-Term Survival*; Harvard University T.H. Chan School of Public Health, Harvard University, September 11, 2012. https://www.hsph.harvard.edu/means-matter/means-matter/survival/

Harvard University, Health Canada. *Medical Assistance in Dying*. Canada.ca. 2018. https://www.canada.ca/en/health-canada/services/medical-assistance-dying.html

Hippocrates. *The Oath by Hippocrates*. The Internet Classics Archive; MIT, n.d. http://classics.mit.edu/Hippocrates/hippooath.html

Lagay, F. Physician-Assisted Suicide: The Law and Professional Ethics. *Virtual Mentor* **2003,** *5* (1). https://doi.org/10.1001/virtualmentor.\pfor1-0301

Lewis, P. J.; Black, I. The Effectiveness of Legal Safeguards in Jurisdictions That Allow Assisted Dying. *SSRN Electr. J.* **2012**. https://doi.org/10.2139/ssrn.2824513

Library of Congress. *Luxembourg: Right to Die with Dignity*. Library of Congress: Washington, DC, n.d. https://www.loc.gov/item/global-legal-monitor/2008-03-02/luxembourg-right-to-die-with-dignity/

Maine Death with Dignity Organization. *Maine Death with Dignity Act Fact Sheet*. Maine Death with Dignity Organization, n.d. https://www.mainedeathwithdignity.org/wp-content/uploads/2020/02/MDWD_FactSheet_2020-02.pdf

Ministry of Health New Zealand. *The End of Life Choice Act 2019*. Ministry of Health NZ, n.d. https://www.health.govt.nz/our-work/life-stages/assisted-dying-service/about-assisted-dying-service/end-life-choice-act-2019

Moss, Z. *Assisted Dying Not Assisted Suicide*. Dignity in Dying, April 10, 2013. https://www.dignityindying.org.uk/blog-post/assisted-dying-not-assisted-suicide/

Nielsen, M. E. J.; Andersen, M. M. Bioethics in Denmark. *Cambridge Q. Healthcare Ethics* **2014,** *23* (3), 326–333. https://doi.org/10.1017/s0963180113000935

Oregon Public Health Division, Center for Health Statistics. *Oregon Death with Dignity Act 2020 Data Summary*. Oregon State Government, 2019. https://www.oregon.gov/oha/ph/providerpartnerresources/evaluationresearch/deathwithdignityact/Documents/year23.pdf

Oregon State Government. *Death with Dignity Act Requirements*; Oregon State Government, n.d. https://www.oregon.gov/oha/ph/providerpartnerresources/evaluationresearch/deathwithdignityact/Documents/requirements.pdf

Palmer, C. *Italy Sees First Assisted Suicide after Years of Court Battles*; Reuters, June 16, 2022. https://www.reuters.com/article/italy-assisted-suicide/italy-sees-first-assisted-suicide-after-years-of-court-battles-idUSL1N2Y30ZZ

Pan American Health Organization. *Parliament of Australia. Euthanasia—The Australian Law in an International Context*. Aph.gov.au; Parliament of Australia, 2019. https://www.aph.gov.au/About_Parliament/Parliamentary_Departments/Parliamentary_Library/pubs/rp/RP9697/97rp4

Pan American Health Organization. *Burden of Neurological Conditions*. www.paho.org; Pan American Health Organization, n.d. https://www.paho.org/en/noncommunicable-diseases-and-mental-health/noncommunicable-diseases-and-mental-health-data-portal-3

Patients Rights Council, PBS News Hour. *The Right to Die in Belgium: An Inside Look at the World's Most Liberal Euthanasia Law*; PBS NewsHour, January 15, 2015. https://www.pbs.org/newshour/show/right-die-belgium-inside-worlds-liberal-euthanasia-laws

Patients Rights Council. *Assisted Suicide Laws in the United States*. Patients Rights Council, n.d. https://www.patientsrightscouncil.org/site/assisted-suicide-state-laws/

Procon.org. *States with Legal Physician-Assisted Suicide*. Euthanasia, July 25, 2019. https://euthanasia.procon.org/states-with-legal-physician-assisted-suicide/

Reuters. *Man Becomes First Person in Colombia with Non-terminal Illness to Die by Legal Euthanasia*; CNN, January 9, 2022. https://www.cnn.com/2022/01/09/americas/colombia-euthanasia-intl/index.html

Riddick, F. A. The Code of Medical Ethics of the American Medical Association. *Ochsner J.* **2003,** *5* (2), 6–10. https://www.ncbi.nlm.nih.gov/pmc/articles/PMC3399321/

Rietjens, J. A. C.; van der Maas, P. J.; Onwuteaka-Philipsen, B. D.; van Delden, J. J. M.; van der Heide, A. Two Decades of Research on Euthanasia from the Netherlands. What Have

We Learnt and What Questions Remain? *J. Bioethical Inquiry* **2009,** *6* (3), 271–283. https://doi.org/10.1007/s11673-009-9172-3

Sperling, D. Travelling to Die: Views, Attitudes and End-of-Life Preferences of Israelis Considering Receiving Aid-in-Dying in Switzerland. *BMC Med. Ethics* **2022,** *23* (1). https://doi.org/10.1186/s12910-022-00785-w

States of Jersey. *Government of Jersey*; Gov.je, n.d. Retrieved June 29, 2022. From https://www.gov.je/Caring/AssistedDying/Pages/AssistedDying.aspx

UCLA Health. *Introduction to California End of Life Option Act*. www.uclahealth.org; UCLA Health, n.d. https://www.uclahealth.org/patients-families/support-information/advance-directive/introduction-california-end-life-option-act#:~:text=The%20California%20End%20of%20Life

Unitarian Universalist Association. *The Right to Die with Dignity*; UUA.org, Unitarian Universalist Association, August 24, 2011. https://www.uua.org/action/statements/right-die-dignity

Verhagen, A. A. E.; Sol, J. J.; Brouwer, O. F.; Sauer, P. J. Deliberate Termination of Life in Newborns in The Netherlands; Review of all 22 Reported Cases between 1997 and 2004. *Nederlands Tijdschrift Voor Geneeskunde* **2005,** *149* (4), 183–188. https://pubmed.ncbi.nlm.nih.gov/15702738/

Vermont Department of Health. *Act 39 Frequently Asked Questions*. Vermont Department of Health, n.d. https://www.healthvermont.gov/sites/default/files/documents/pdf/Act39_faq.pdf

Vizcarrondo, F. E. Neonatal Euthanasia: The Groningen Protocol. *Linacre Q.* **2014,** *81* (4), 388–392. https://doi.org/10.1179/0024363914z.00000000086

Washington Government. *The Washington Death with Dignity Act*. wa.gov; Washington Government, 2009. https://apps.leg.wa.gov/RCW/default.aspx?cite=70.245&full=true

World Medical Association. *WMA International Code of Medical Ethics*; World Medical Association, April 28, 2010. https://web.archive.org/web/20100428131219

World Medical Association. *WMA Statement on Physician-Assisted Suicide*. World Medical Association; World Medical Association, July 25, 2014. https://web.archive.org/web/20140725091434

World Medical Association. *WMA Declaration of Geneva*; World Medical Association, 2017. https://www.wma.net/policies-post/wma-declaration-of-geneva/

WorldData.info. *Catholics—Worldwide Distribution*; WorldData.info, n.d. https://www.worlddata.info/religions/catholics.php#:~:text=The%20Catholic%20churches%20form%20a

Worley, H. *The Growing Global Chronic Disease Epidemic*; Population Reference Bureau, May 17, 2007. https://www.prb.org/resources/the-growing-global-chronic-disease-epidemic/

CHAPTER 10

THE GROWING WORLD OF BIRTH TOURISM

ABSTRACT

Birth tourism is a controversial topic in the medical tourism industry and medical world at large, especially when considering its larger effects on international countries' immigration policies. From anchor babies to bad-faith birth tourism agencies, this chapter explores the root causes of why foreigners may participate in birth tourism, the response of international media and politicians to the practice, nations' evolving policies on birthright citizenship as direct response to birth tourism, and potential solutions to the practice.

While we have previously explored the reasoning behind patients turning toward medical tourism and travel for treatment for ways to end their lives with dignity, the global health industry is likewise filled with controversies surrounding the other side of the same coin: birth tourism. Theoretically simple in its premise, birth tourism refers to a person from one country traveling to another to give birth, with the end goal of obtaining birthright citizenship—also known as *jus soli*—and passports for their child from their destination of choice. However, birth tourism in practice is considered widely controversial across the world and tinged with issues like xenophobia and racism, leading many countries to end their *jus soli* policies and adopt more restrictive immigration policies.

The ability to participate in birth tourism varies from country to country and is dependent on each individual nation's legislation. Countries with unrestricted *jus soli* in 2022 include the United States, Canada, Mexico, Antigua and Barbuda, Barbados, Belize, Costa Rica, Cuba, Dominica, Grenada,

Medical, Dental, and Wellness Tourism: A Postpandemic Perspective. Mary Schreiber Swenson, PhD, and Amit Bansal (Authors).

Guatemala, Honduras, Jamaica, Nicaragua, Panama, Saint Kitts and Nevis, Saint Lucia, and Trinidad and Tobago in North America; Argentina, Bolivia, Brazil, Chile, Ecuador, Guyana, Paraguay, Peru, Uruguay, and Venezuela in South America; Chad, Lethoso, and Tanzania in Africa; and Pakistan in Asia.

10.1 CONTROVERSIES

The concept of "birth tourism" is inherently controversial, as many opponents equate the practice as an easy way for foreigners to gain access to another country without properly immigrating; however, opponents of birth tourism lump in foreigners who have had an established legal presence in a country, close ties to their community, and the intention to formally emigrate ahead of pregnancy as the same as traveler's who arrive for the explicit purpose of birth tourism, which can contribute to an increase in xenophobia against foreign residents. This is particularly visible in countries like the United States, where the topic of immigration has become a major voting issue and stratifier of congressional party lines—an ironic stance, considering the United States was founded as a "melting pot" of different cultures.

"Birth tourism" discourse in the United States is heavily racist in its rhetoric. Acting US representatives have referred to pregnant migrants entering the country with offensive and derogatory phrases that include "dropping babies", "rats [who] multiply", "anchor babies", "invasion by birth canal", and "terror babies". While running as a presidential candidate, Jeb Bush infamously declared that his use of the term 'anchor baby' was not directed at the Hispanic population, but "frankly . . . Asian people". He is certainly not an outlier. One study found that one of the defining characteristics of birthright citizenship discourse amongst US Republican candidates was the targeting of Latina and Asian women as threats to American society and values. The racialisation of "birth tourism" conforms to a long pattern of criticizing racialised women for 'unethical' reproductive decisions, often the idea that women have children to gain state resources. (Chudleigh, 2021)

Many birth tourism critics in the United States refer to the practice as being "illegal," although birth tourism is completely legal under US law. According to the US Constitution's Fourteenth Amendment, which states that "all persons born or naturalized in the United States, and subject to the jurisdiction thereof, are citizens of the United States and of the State wherein they reside" (National Archives, 2018). Despite this, US politicians have attempted to restrict or altogether ban the practice in the United States, with some calling for amendments to the constitution to put a permanent end to US *jus soli*.

In 2020, President Donald J. Trump's administration announced a new policy that would see the State Department "no longer issue temporary visitor (B-1/B-2) visas to aliens seeking to enter the United States for 'birth tourism.'" However, this provision did not apply to the 39 countries—mostly European—who are a part of the US Visa Waiver Program, raising concerns about what races and ethnicities were specifically targeted by the new policies, as well concerns about how the policy would be implemented due to the fact that consular offices cannot directly ask visa applicants if they are pregnant (Atwood et al., 2020).

> Unlike the argument for protection of medical resources, American racism is reflected in the year-old ruling. The ban on pregnant travellers on B-visas does not extend to those from the 38 countries in the Visa Waiver Program, most of which are majority-white and European. The restrictions still allow for pregnant visitors from these states, affecting primarily non-white visitors....
>
> ....While some may justify the ruling on the grounds that the U.S. has the right to patrol its own borders, this ban actively harms the mobility and agency of vulnerable migrants. Not all so-called "birth tourists" enter the United States in search of citizenship for their child, as the ruling suggests. Some mothers travel for health care that may otherwise be unavailable to them. The 2020 restrictions have no exemptions for those who seek American healthcare for their own safety and wellbeing. A State Department official commented on these mothers in an anonymous interview: "People will die because of this." (Chudleigh, 2021)

Though many opponents of birth tourism point to the Trump-era policy update as a preservation of American resources for the good of the American policy, the structure of the policy seems to directly contradict that ideal; if foreigners are stealing Americans' access to hospital beds and US medical care, why is the policy only extended to birth tourists, and not foreigners looking to enter the United States for any other kind of medical procedure? According to Chudleigh, "if the outrage was a matter of medical resources, public debates would extend to all medical tourists, not just pregnant ones." Not to mention, the United States is one of the only two countries in the world that taxes its citizens on their income no matter where they live in the world; subsequently, children born to birth tourists in the United States will theoretically be contributing to the United States economy and tax reserve regardless of if they actually live in the United States—or if they have even set foot on US soil—putting a dent in the idea that birth tourism is a strain on American resources (US Internal Revenue Service, n.d.).

The racial rhetoric seems to pervade deeper into US politicians' stances on the matter, with some claiming that birth tourism is a form of "cheating" one's way into the United States and that it damages the sanctity of American citizenship. Considering the United States is a colonized country founded and built on the backs of immigrants—not to mention the widespread racial injustices committed throughout the country's formative years, especially through slavery—such points of view can be considered tinged with racism and xenophobia.

> There have also been several suggestions put forward by members of Congress including the Stop Birth Tourism Act of 2015 by Representative Rohrabacher of California, sought to amend the Immigration and Nationality Act by making certain pregnant nonimmigrant aliens ineligible for visas or admittance into the United States, but failed to pass in committee. The Act would have enabled consular officers to require certification from a healthcare provider that the alien is not pregnant and it would have made undocumented immigrants who were found to be pregnant, and likely to give birth, inadmissible. This bill sought to address birth tourism by making it more difficult for potential parents to obtain visas or enter the United States, rather than to alter birthright citizenship for their children. President Trump has proposed more radical measures to discourage unauthorized immigration that would also limit birth tourism by changing the interpretation of the 14th Amendment to exclude birthright citizenship for individuals born to two non-citizens. (Kugler, 2018)

Some US politicians, however, have spoken out in favor of upholding the constitution on the subject of birth tourism, though not going so far as to advocate for the practice itself. "We believe in a clean slate principle," said Clinton-era Assistant Attorney General Walter Dellinger. "Whatever questions there are about the legitimacy of parents or grandparents, in our country you get a clean slate. Every new child who is born here is simply and indisputably an American. And that is part of our almost unique national identity" (Neal, 2010).

Another large controversy surrounding birth tourism in the United States has been the existence of birth tourism groups, which help set up arrangements for pregnant foreigners to give birth inside the United States. While high-income tourists can afford high-end accommodations and a properly trained physician, some of these groups target low-income women and put them in situations medically unfit or unsafe for the birth of a child—a trend that clearly demonstrates that not all birth tourists are pulling away resources from the American people, as postured by opponents.

According to the 2018 investigation by the New York Times, many daycare centers across the United States are actually facilities for the birth of pregnant foreigner's children, using the daycare status as a way to avoid questions by authorities. This puts the mothers at high risk for complications due to a lack of professional physicians, and can likewise put the baby's life in jeopardy. Still, for many low-income women, insecure resources such as these are their only viable opportunity to have a child born in the United States, particularly in the post-Trump policy era (Goldbaum, 2018).

To former US consular official Christopher Richardson, the Trump anti-birth tourism policy does little to solve the problem at its root; "instead of cracking down on consular officers, the United States should crack down on the birth tourism industry. A big part of the problem could be addressed by identifying hospitals and other medical facilities that profit from the foreign-national birth business and impose prohibitive fines on them if they continue to engage in the practice. These facilities often work hand in glove with birth tourism consultants, so this would help suppress that element as well" (Richardson, 2020).

While the controversies over birth tourism are perhaps most visible in the United States, it's not the only country wielding a seemingly racist rhetoric to attack the problem.

> We must recognize that, though racist controversies over "birth tourism" are the loudest in the United States, they are not unique to it. In 2018, the Canadian Conservative Party passed a non-binding resolution to end birthright citizenship due to concerns over rising "birth tourism" at their annual convention. Ireland operated under birthright citizenship until a constitutional amendment in 2004, one that was said to be motivated by the increases in immigration and 'baby tourism'. Both of them were home to similar, racially-charged language in their public "birth tourism" debates. (Chudleigh, 2021)

Outside of the United States, Canada has become one of the most popular destinations for birth tourism in recent years, with birth tourism reportedly increasing by 13% year over year from 2018 to 2019 according to data from the Canadian Institute for Health Information (Favaro, 2019), which similarly revealed that 1.4% of all births in Canada during 2019 were from nonresidents. Still, the true number of birth tourists in Canada is difficult to ascertain as Canada's "hospital coding used to capture uninsured patients does not selectively capture birth tourists" and may use other coding, such as "Uninsured Prenatal Patient", when creating a pregnant nonresident patient's file (Brar et al., 2022).

There is a number of reasons birth tourists have increasingly turned to Canada as a birth tourism destination; the country is well known for its robust public healthcare system and offering significant discounts on education to its residents, amongst other benefits. According to a research study of Canada's birth tourism, "77% of Birth Tourists stated that their primary reason to deliver their baby in Canada was for the desire for a Canadian baby who would be eligible for Canadian citizenship" (Bahr).

Much like in the United States, Canada's growing birth tourism industry has caused repeated controversies throughout the country, including some instances of racial rhetoric and an evident divide in public opinion. Studies show that "concerns about maternity tourism are synchronous with anxieties around the impact of foreign (read: Asian) investors driving up real estate prices and creating an affordability crisis in housing markets," showing the potential racial and xenophobic roots of Canadian anti-birth tourism stances (Harder, 2020).

On the political end of the spectrum, numerous Canadian politicians have spoken out against birth tourism and in favor of amending Canada's nationality law. Parliament member Joe Peschisolido publicly called the practice "unethical and unscrupulous," endorsing a petition to end birth tourism in Canada and claiming that birth tourism "can be used to gain access to Canada's publicly subsidized post-secondary education system and to take advantage of Canada's public healthcare system and generous social security programs, all without having to contribute much to the funding of these systems and programs" (Bilefsky, 2018; Jung and Richmond, 2020).

Advocates more sympathetic to Canadian birth tourism, like the British Columbia Civil Liberties Association, have referred to politician's claims of a drain on Canada's resources as unfounded, and pointed out that the benefits of being a Canadian citizen are only applicable to people who actually live in Canada; taking into consideration that nonresidents are not allowed to stay in Canada simply because their baby was born there, advocates say that most nonresident birth tourists must leave the country with their baby in tow after their six-month tourist visa has expired.

> Eliminating birthright citizenship would impose enormous and perpetual public expense that will be borne by all of us. Currently, a Canadian birth

> certificate serves as proof of citizenship for birthright citizens, whether for obtaining a passport, or registering as a voter. But if being born in Canada does not automatically confer citizenship, then the provinces and federal government will eventually have to come up with another system for verifying citizenship. A passport proves citizenship, but the government makes us pay for them and about a third of Canadians don't possess one. Whatever the verification system, it will be staggeringly expensive to develop and maintain. By way of example, a national identification card system—which is itself fraught with privacy and surveillance concerns—has been estimated to cost between $3 billion to $5 billion just to implement. (Cheung, 2014)

10.2 EXPLORING THE MOTIVATIONS BEHIND BIRTH TOURISM

An integral part of understanding the reasons behind birth tourism is delving into *where* these birth tourists originate from and *why* they are seeking citizenship for their child elsewhere in the world, though exploring the data is no easy feat considering most birth tourism is under the table and clandestine. Still, with media reports of approximately 36,000 birth tourists entering the United States per year, there's enough information out there to glean certain insights about the international birth tourism industry at large (Camarota, 2015).

> The origins of modern birth tourism can be linked to several "push and pull factors" in the origin and receiving countries. Pull factors include political stability or economic opportunities in the receiving country, and push factors include economic uncertainty or political risk in the country of origin. Push factors refer to "the reasons why people leave an area" whereas "pull factors" reference "the reasons why people move to a particular area." Many parents choose to give birth in the United States because of the ease of birthright citizenship. However, other pull factors include high quality paternity care, a strong economy with a robust job market, relative political stability, and a leading public education system. As to push factors, many parents want to give their children the option of a life in the United States because of their fears surrounding the stability of their own countries, the potentially limited opportunities for social advancement, disparities in the educational systems, and economic concerns. (Kugler, 2018)

Some birth tourists look to give birth abroad to have an "anchor baby"—a contentious term referring to a child's ties to a country due to their citizenship. For some tourists, this means having their child in a place like the United

States, returning with child-in-hand to their home country, and returning to the United States when the child is 21 and legally allowed to help blood relatives immigrate into the country. For others, the "anchor baby" concept will see them remain in the United States or the child's respective country of birth in hopes that having a citizen as a child will help their efforts to naturalize or legally immigrate into said country.

Though the "anchor baby" argument is widely used to antibirth tourism advocates, there is a myriad of other factors that can come into a play among a traveler's decision to give birth abroad, including political unrest, access to medical care, gender roles, educational opportunities, and restrictions on how many children one household can have, just to name a few. To characterize all birth tourism as a way to obtain citizenship would be to overlook the important intersectional nuances that lead future parents into making these choices for themselves and their children.

Take China, for example, a country that had a restrictive one-child birth policy as a form of population control from 1980 up until 2016, when it was updated to allow up to two children per family unit (Wang et al., 2016). As a result of these policies, experts estimate that the large majority of birth tourism comes from Chinese nationals, but due to the lack of data on the largely clandestine practice, it's difficult to surmise precisely how many birth tourists originate from the world's most populated country. Still, according to an investigative report, there were more than 500 birth tourism agencies operating and advertising their services in China as of 2015—one year before the one-child policy was ended—demonstrating the practice's clear popularity in the nation (MacLeod, 2015).

While the root source of Chinese birth tourism has been largely framed by Western media as foreigners looking to take advantage of US resources, analyses such as these often overlook China's restrictive birth policies. For many Chinese nationals, birthing a child abroad was one of the few ways to avoid the government's one-child policy regulations, as foreign-born children are not considered privy to the policy if they do not possess Chinese citizenship. The United States is not the only country to receive an influx of Chinese birth tourists; a reported 200,000 babies were born to Chinese nationals in Hong Kong between 1979 and 2015 (Chee, 2021).

An examination of Western attitudes toward Chinese birth tourism shows repeated claims that expecting Chinese women flock to the United States solely because they wish their child to be an American citizen (MacLeod, 2015). While some tourists likely have intentions such as these, it completely overlooks and disvalues the pain the one-child policy caused countless

Chinese families, many of which terminated their pregnancies of female fetuses in hopes of later becoming pregnant with a male due to China's historic preference toward men. Violators of the one-child policy could be with massive fines, forced abortions, loss of employment, and, in some cases, full-on sterilization (Mullen, 2021). As a result, a number of Chinese birth tourists may have participated in the industry simply to avoid terminating a wanted pregnancy, or for the chance to expand their family without falling victim to harsh punishments—especially considering the mental anguish experienced during the one-child policy that still lingers throughout China to this day (Feng, 2021).

Now that China revised its one-child policy in fall 2021 to allow families up to three children per household due to the country's declining birth rate, it can be anticipated that outbound Chinese birth tourism will decline in turn (BBC News, 2021). But as biases against Chinese birth tourists linger in main birth tourism destinations like the United States, it's important to read between the lines and understand *why* there have been so many Chinese birth tourists with additional empathy and understanding in mind.

Korean nationals have likewise caught flack for participating in birth tourism in the United States, and numerous clandestine facilities catering to pregnant Korean women have been shut down in Los Angeles in recent years (Han, 2013), while Korean nationals reportedly make up the majority of foreigner births in the US territory of Guam (Cruz, n.d.). Many of these births are indeed for the purpose of obtaining US citizenship for their children and all the benefits that come with US nationality, though these are anticipated to decline following Korea's choice to begin offering its citizens financial incentives to birth their children in Korea to help bolster the country's birth rate as of 2020 (Sajid, 2020).

For Korean birth tourists, much of the drive toward birth tourism can be attributed to Korea's vast economic inequality, where the cost of living is high and the wages can be low. Considering the cultural pressure to provide their children with an elite education, many Korean parents reportedly spend large portions of their income on their child's schooling, which can cause undue economic stress on the family (Deutsche Welle, 2021). Birth tourism provides opportunities for Korean nationals to give their children the educational and financial resources *jus soli* countries like the United States offer their citizens, and perhaps one of the viable opportunities to raise a thriving family without being restricted by Korea's economic barriers.

Nigeria, too, has become a major source of incoming birth tourists to the United States and other countries, with many looking to give their children the opportunities that come with citizenship abroad. The root causes of Nigerian birth tourism are clear: 40% of the population lives below the poverty line, a fate many Nigerian parents hope to eliminate from their child's future through birth tourism, though the majority of Nigerian birth tourists to the United States are reportedly mainly professional and middle-class Nigerians (World Bank, 2020).

The United States has likewise received an influx of Russian-birth tourists, with many of these Russian nationals specifically choosing Florida as their destination of choice. According to reports, Florida's hot climate and American friendliness stand in stark contrast to Russia's colder nature and turn birth tourism into what feels like a vacation, maximizing travel for treatment to the fullest (Stashevska, 2019). Russians have similarly been attracted to the South Florida's metropolis of Miami for its bustling Russian population, finding a slice of familiarity abroad. Many have been prompted to travel to the United States due to memories of Soviet-era rules and the continued geopolitical conflict in Eastern Europe, hoping to give their children the chance to avoid these very same issues as they grow up.

> Additionally, perceptions of life in the United States are magnified by pop culture in other countries, such as the Chinese film "Finding Mr. Right" which romanticized the experiences of a woman who comes to give birth in the United States. Furthermore, many are enticed by the alluring pictures painted by the companies engaged in the practice. The "AmeriMama" program run by a New Jersey hospital marketed birthing services costing between $8,500 and $27,500 to Russian mothers on the premises of their children receiving U.S. citizenship. While some of these "birth tourism" operators run safe programs, others aggressively cut corners resulting in higher death rates and complications for children of "birth tourists." (Kugler, 2018)

For the majority of birth tourists, birthing a child in another country will give said child access to healthcare, education, financial and professional opportunities, and above all, security that they would not be privy to in their parents' countries of origin, for a price tag of tens of thousands of dollars, when sourced from birth tourism agencies. While each individual cases' motivations may vary, the common thread throughout seems to be the desire to give a child a better chance at life—and considering most US birth-tourism opponents are on the right side of the aisle, and subsequently, likely prolife, perhaps antibirth tourism advocates may find that certain aspects of birth tourism aligns with their ideologies upon additional analysis.

That being said, the birth tourism industry at large clearly needs sweeping reforms, but without addressing the root causes in tourists' respective countries of origin, clandestine birth tourism is likely to continue in some capacity.

10.3 COUNTRIES' EVOLVING POLICIES TO COMBAT BIRTH TOURISM

In an effort to curb birth tourism, numerous nations across the world have made legislative and policy changes to end the concept of *jus soli* and make obtaining citizenship decidedly more difficult. Instead, many have adopted a *jus sanguinis* approach instead, which only grants citizenship to a child born in a country if at least one (or both) of a child's parent is a citizen of said nation in question (Giacomin, 2020).

The United Kingdom did not officially codify its citizenship law until 1914's British Nationality and Status of Aliens Act 1914, which granted immediate British citizenship to anyone born within the nation (Legislation.gov.uk, n.d.). However, growing unrest surrounding the United Kingdom's increasing immigration rate in the 1960s and 1970s—especially nonwhite migrants—caused the country to tighten its immigration policies and end its previous *jus soli* provisions. The introduction of the British Nationality Act 1981 made *jus sanguinis* the new standard of obtaining British nationality at birth, requiring at least one parent to be a British national or legal permanent resident of the nation—criteria most birth tourists fail to meet (Legislation.gov.uk, 2010).

The United Kingdom's efforts to curb birth tourism have resulted in continued issues for the nation's immigration policy, as the 1981 protective provision that extended citizenship to children born in the United Kingdom to nonpermanent residents or citizens after the age of 10 faced continual changes in the years following, including the addition of exorbitant fees, and made it increasingly difficult to obtain British citizenship.

> The basis for that expectation has been radically altered by relatively recent changes to immigration rules and policy, which have greatly extended the period before someone with lengthy residence in the UK can apply for settlement. In making those changes, no regard was had to Parliament's expectation and intention in passing the 1981 Act. The effect of this is dramatic and harmful. Many children, whom Parliament expected to be born British, are not born British. This is because the child's parents are not settled at the time of her or his birth, despite having lived in the UK for long

> periods that Parliament had expected would lead to their being settled. This includes parents who previously would have become settled during their own childhoods in the UK. (Valdez-Symonds, 2018)

Ireland implemented a similar policy in 2004 after years of allowing *jus soli*, amending its constitution with the Twenty-seventh Amendment of the Constitution Act 2004, which states that "a person born in the island of Ireland, which includes its islands and seas, who does not have, at the time of the birth of that person, at least one parent who is an Irish citizen or entitled to be an Irish citizen is not entitled to Irish citizenship or nationality, unless provided for by law" (Electronic Irish Statute Book (eISB), n.d.). This was prompted by a number of factors; in 2002, 15% of Ireland's record number of births were reported as births to non-nationals; former Irish Minister for Justice Michael McDowell similarly claimed that 40–50% of births from non-nationals in Ireland were made for the explicit purpose of obtaining the children Irish citizenship, though he failed to provide the relevant data leading to his figures (RTÉ Ireland, 2004). The amendment was ultimately approved in a public referendum, with nearly 80% of Irish voters voting in its favor (RTÉ Archives, n.d.).

While previously granting constitutional right citizenship to anyone born within its borders, India went on to amend its birthright citizenship numerous times in the future in an attempt to curb its rapidly growing population (Ministry of Home Affairs Government of India, n.d.). India's Citizenship Amendment Act, which went into effect in 2004, continued to grand *jus soli* citizenship to anyone born in India between January 26, 1950, and July 1, 1987, and changed requisites for those born between July 1, 1987, and December 3, 2004, to need at least one Indian national parent in order to obtain citizenship. This legislation likewise made illegal immigrants personaae non gratae in the eyes of immigration, making undocumented migrants ineligible for both citizenship and pathways to naturalization (PRS Legislative Research, 2019). Children of undocumented immigrants were affected, too; any child who has at least one parent that is living illegally in India is automatically disallowed from obtaining Indian citizenship, essentially putting an end to birth tourism in the country.

India subsequently implemented additional changes to its nationality law in 2016 with the Citizenship Amendment Bill, which went into effect in 2019 (Hausman, 2020). This allowed members of certain religions who immigrated from certain countries—Buddhists, Christians, Hindus, Jains, Parsis, and Sikhs from Afghanistan, Bangladesh, and Pakistan—who immigrated to

India before 2014 to no longer be considered illegal immigrants and granted them easier opportunities to achieve Indian citizenship. Despite opening up its immigration policy to a certain degree, India's 2016 amendment caught flack for excluding Muslims and Jews immigrants from these provisions, which has been characterized as discriminatory and as a "Nuremberg moment" (Mahmudabad, 2019). Proponents of the Indian law have dismissed claims of discrimination, saying that "it is meant for non-citizens residing in India to escape religious persecution in three religious states bordering India by providing citizenship" (Bhat et al., 2020).

These are far from the only countries across the world to rethink their *jus soli* status; Australia, Germany, France, Pakistan, South Africa, New Zealand, Malta, and the Dominican Republic have all made amendments to their nationality laws to restrict birth tourism and birthright citizenship. Meanwhile, the United States and Canada, who have maintained *jus soli*, continue to be two of the world's most popular birth tourism destinations.

10.4 THE IMPLICATIONS OF BIRTH TOURISM ON THE MEDICAL TOURISM INDUSTRY

Though birth tourism may fall in line with the definition of medical tourism on paper—that is, traveling to receive medical care in another destination due to financial or quality incentives—the actual practice of birth tourism itself is not indicative of medical tourism and travel for treatment as a whole, and birth tourism agencies as a whole should not be lumped in with the rest of the medical tourism industry. Considering many birth tourism agencies will actively encourage their clients to lie at customs and immigration about their reason for travel and to hide their pregnancies, which can be constituted as visa fraud, the birth tourism sector can have reverberating negative effects on medical tourism as a whole (Phillip, 2015).

Take the aforementioned maternity hotels, for example. While the medical tourism industry is intent on providing its clients high-quality care for cost-efficient prices, the birth tourism industry itself often funnels its clients into clandestine maternity hotels that are missing the medical professionals and adequate healthcare needed to ensure a safe and secure birthing process. Though this lack of medical security mainly affects lower income birth tourists, the lack of ensured regulation and safety is a far cry from the strict standards the medical tourism industry holds itself to across the globe.

The wealthier birth tourists seem to fare better than their lower income counterparts in terms of quality of care, and are offered cushier stays—oftentimes in hotels or higher end clandestine facilities—for a price tag approximated between $20,000 and $80,000.

> Women who book rooms at these properties can expect to live in well-stocked apartment complexes or large suburban homes with laundry and catered food as part of the package. Once their babies are born in an American hospital, they are cared for by nurses while the mothers rest for at least a month. They can pass their time with shopping trips to luxury stores, trips to amusement parks or poolside at the "hotel" while attentive caretakers look after the infants, feeding, bathing and putting them to sleep on a regimented schedule. (Phillip, 2015)

While there's clearly plenty of money to be had in the birth tourism industry, most birth tourism agencies prefer to operate strictly in cash to avoid declaring said income and shirking tax requirements, with many birth tourists paying a deposit up front and later paying the rest of their duty upon arrival. One Southern California maternity hotel reportedly brought in more than $1 million in undeclared income in a single year before being busted by federal agents in 2013.

With their cash already in hand, some birth tourism agencies go on to encourage their clients to "scam" the US hospital system for additional savings.

> As women went into birth, they were taken to local hospitals and declared jobless. As many as 400 babies associated with just one of these companies were born after 2013 in Orange County, Calif., hospitals. Despite the fact that many of these women paid tens of thousands of dollars to participate in the "maternity hotel" scheme, they claimed to be unable to pay the hospitals, which typically charged about $25,000 per birth. (Phillip, 2015)

Continued crackdowns on these facilities are expected to culminate in "the biggest federal criminal case ever" against the birth tourism industry in the United States and are anticipated to surround charges of tax evasion, visa fraud, and harboring illegal immigrants (Jordan, 2015).

Though birth tourism itself is not illegal, and birth tourists' intentions in participating in the practice are generally for the good of the child, birth tourism agencies continue to prey on foreign women and encourage them to commit fraud and enter the United States under false pretenses—clear violations of the law. However, this should not reflect on the medical tourism industry as a whole, which generally works in collaboration with countries' tourism offices, accredited institutions, and follows procedure to the letter of international law for the good of its clients and the integrity of the industry.

That being said, if the birth tourism industry is overhauled and reformed, there could be a place in the medical tourism industry for controlled and legalized birth tourism, especially as "birth tourism can also serve as an engine for local economies, with parents spending money at local stores on general products, medical services, or legal help" (Kugler, 2018).

10.5 POTENTIAL SOLUTIONS TO BIRTH TOURISM

Birth tourism is an undeniably complex subject that cannot be solved overnight in countries like the United States and Canada, especially when considering that birthright citizenship is a fundamental aspect to their respective constitutions.

In the United States, continued crackdowns on maternity hospitals and birth tourism agencies could help remove unsafe and insecure facilities out of the equation, subsequently providing better care and outcomes for birth tourists—even if the practice is unregulated. Currently, most crackdowns are on so-called nuisance maternity hotels that disturb the residents around them; according to Kugler, "this practice could be expanded beyond targeting 'nuisance' operators to targeting less reputable facilities that may endanger the health and wellbeing of the mother and child, on health code grounds. This crackdown may be done in conjunction with ICE by targeting operators assisting in visa fraud and tax evasion in violation of federal immigration laws."

> U.S. embassies operating in countries with significant numbers of birth tourists such as Brazil, China, Mexico, Nigeria, Russia, South Korea, and Turkey should inform visa applicants of the consequences of visa fraud and the risks of working with unregulated operators. These actions could be advanced by a collaboration between local and federal authorities, focusing on the more crooked organizations that operate in this industry. Narrowing the focus onto these facilities may also be more feasible than a major policy change, and may be a more efficient use of government resources since these facilities do not provide adequate care for their patients....

> ….A more targeted approach would likely address the limited damages of the practice, while not constituting an unnecessarily expansive and legally complicated response such as restricting birthright citizenship. (Kugler, 2018)

In Canada, certain regions have already implemented their own anti-birth tourism measures, as the city of Calgary, Alberta added an upfront $15,000 CAD deposit for prenatal delivery and post-natal care over summer 2019 for all nonresidents (Labby, 2022). This additional monetary requirement could potentially discourage birth tourists from birthing their children in Calgary; however, given this policy's implementation less than a year before the pandemic and its subsequent widespread travel restrictions, there's not enough data yet to see if this policy has discouraged birth tourism to Calgary, especially as nonresident births dropped 57% during the pandemic (Griffith, 2021).

Canada could potentially face legislative changes to its nationality law, particularly as a survey conducted by the Angus Reid Institute in 2019 revealed that 64% of Canadians think that a child born to parents on a tourist visa should not be automatically given Canadian citizenship, while another 60% said that Canada needs to change its citizenship laws accordingly to end the practice of birth tourism in the country. That being said, some 40% of Canadians said they believe retaining *jus soli* in Canada is a positive policy, compared to 33% who perceived it as negative (Angus Reid Institute, 2019), which could affect a potential referendum down the line if that's the course Canada chooses to take. However, as pointed out by the British Columbia Civil Liberties Association, such legislation changes would be costly to the Canadian public.

There's no simple solution to birth tourism, especially as public opinion across the world remains divided on the subject. But with careful understanding of birth tourists' motivations and backgrounds, increasing crackdowns on unsafe clandestine facilities and birth tourism agencies, and creating an open dialogue with over countries to discourage their citizens from participating in birth tourism, it is possible to potentially reform the birth tourism sector at its core without having to amend constitutions.

KEYWORDS

- **medical tourism**
- **travel for treatment**
- **birth tourism**
- **anchor babies**
- ***jus soli***
- ***jus sanguinis***
- **immigration policy**

REFERENCES

Angus Reid Institute. *Birthright Citizenship: Plurality of Canadians See It as Good Policy, But Also Say Some Changes Are Needed*; Angus Reid Institute, Mar. 14, 2019. angusreid.org/birthright-citizenship-birth-tourism/ (accessed July 14, 2022).

Atwood, K.; et al. US Issues New Rules Restricting Travel by Pregnant Foreigners, Fearing the Use of "Birth Tourism." *CNN*, Jan. 24, 2020. edition.cnn.com/2020/01/23/politics/us-new-rules-restricting-travel-fearing-birth-tourism/index.html (accessed July 4, 2022).

BBC News. *China NPC: Three-Child Policy Formally Passed into Law*; BBC News, Aug. 20, 2021. www.bbc.com/news/world-asia-china-58277473

Bhat, S.; et al. Misplaced Outrage on India's New Citizenship Laws: Opinions of Healthcare Professionals. *J. Fam. Med. Primary Care* **2020,** *9* (5), 2188. doi:10.4103/jfmpc.jfmpc_245_20 (accessed Aug. 8, 2020).

Bilefsky, D. "Birth Tourism" Is Legal in Canada. A Lawmaker Calls It Unscrupulous. *The New York Times*, Dec. 31, 2018. www.nytimes.com/2018/12/31/world/canada/china-birth-tourism.html

Brar, S.; et al. Impact of Birth Tourism on Health Care Systems in Calgary, Alberta. *BMC Health Serv. Res.* **2022,** *22* (1). doi:10.1186/s12913-022-07522-4 (accessed Mar. 19, 2022).

Camarota, S. *There Are Possibly 36,000 Birth Tourists Annually*. CIS.org, Center for Immigration Studies, May 28, 2015. cis.org/Camarota/There-Are-Possibly-36000-Birth-Tourists-Annually (accessed July 4, 2022).

Chee, W.-C. How Family Policies Redefine Families: The Case of Mainland China-Hong Kong Cross-Border Families. *Int. J. Soc. Welfare* **2021,** *30* (4), 478–489. doi:10.1111/ijsw.12491.

Cheung, C. *Born Equal: Citizenship by Birth Is Canada's Valuable Legacy*; BC Civil Liberties Association, Aug. 18, 2014. bccla.org/2014/08/born-equal-citizenship-by-birth-is-who-we-are/ (accessed July 14, 2022).

Chudleigh, S. Race over Resources: The American Controversy over "Birth Tourism." *Social Policy*; London School of Economics and Political Science, Mar. 22, 2021. blogs.lse.

ac.uk/socialpolicy/2021/03/22/race-over-resources-the-american-controversy-over-birth-tourism/ (accessed July 4, 2022).

Cruz, M. Maternity Tourism Rates Consistent for Korean Visitors. *The Guam Daily Post*, n.d. www.postguam.com/news/local/maternity-tourism-rates-consistent-for-korean-visitors/article_2295000a-5d61-11e7-822e-0bb475a77054.html (accessed July 6, 2022).

Deutsche Welle. *What's behind South Korea's Population Decline?* December 21, 2021. dw.com. www.dw.com/en/whats-behind-south-koreas-population-decline/a-60210895

Electronic Irish Statute Book (eISB). *Twenty-Seventh Amendment of the Constitution Act, 2004*. Electronic Irish Statute Book (eISB), n.d. www.irishstatutebook.ie/eli/2004/ca/27/enacted/en/html (accessed July 12, 2022).

Favaro, A. "Birth Tourism" Rising Fast in Canada; Up 13 Per Cent in One Year. *CTVNews*, Sept. 16, 2019. www.ctvnews.ca/health/birth-tourism-rising-fast-in-canada-up-13-per-cent-in-one-year-1.4594970?cache=yesclipId104062 (accessed July 4, 2022).

Feng, E. *China's Former 1-Child Policy Continues to Haunt Families*; NPR.org, July 4, 2021. www.npr.org/2021/06/21/1008656293/the-legacy-of-the-lasting-effects-of-chinas-1-child-policy

Giacomin, N. Jus Soli and Jus Sanguinis: Understanding the Two Concepts. *Space Legal Issues*. Feb. 25, 2020. www.spacelegalissues.com/understanding-jus-soli-and-jus-sanguinis/

Goldbaum, C. It Was a Suspected "Birth Tourism" Site. Then Came Late-Night Screams. *The New York Times*, Sept. 21, 2018. www.nytimes.com/2018/09/21/nyregion/queens-stabbing-day-care-birth-tourism.html?action=click&module=In%20Other%20News&pgtype=Homepage&action=click&module=News&pgtype=Homepage (accessed July 4, 2022).

Griffith, A. Birth Tourism in Canada Dropped Sharply Once the Pandemic Began. *Policy Options*, Dec. 16, 2021. policyoptions.irpp.org/magazines/december-2021/birth-tourism-in-canada-dropped-sharply-once-the-pandemic-began/ (accessed July 10, 2022).

Han, J. LA Targets Korean "Birth Tourists." *Koreatimes*, Feb. 4, 2013. koreatimes.co.kr/www/news/opinion/2013/02/137_130010.html (accessed July 6, 2022).

Harder, L. "Maternity Tourism," Civic Integration and Jus Soli Citizenship in Canada. *Rev. Eur. Migr. Int.* **2020,** *364* (4), 35–54. www.cairn.info/revue-europeenne-des-migrations-internationales-2020-4-page-35.htm (accessed July 10, 2022).

Hausman, G. Citizenship Amendment Act (CAA) and National Register of Citizens (NRC)—Global Studies Blog. *Blogs.cul.columbia.edu*, Dec. 10, 2020. blogs.cul.columbia.edu/global-studies/2020/12/10/citizenship-amendment-act-caa-and-national-register-of-citizens-nrc (accessed July 10, 2022).

Jordan, M. Federal Agents Raid Alleged "Maternity Tourism" Businesses Catering to Chinese. *Wall Street Journal*, Mar. 3, 2015. www.wsj.com/articles/us-agents-raid-alleged-maternity-tourism-anchor-baby-businesses-catering-to-chinese-1425404456?tesla=y (accessed July 10, 2022).

Jung, A.; Richmond, B. C. Politicians Push Ottawa to Address Birth Tourism and Stop "Passport Mill." *British Columbia*, Sept. 5, 2020. bc.ctvnews.ca/richmond-b-c-politicians-push-ottawa-to-address-birth-tourism-and-stop-passport-mill-1.5094237 (accessed July 10, 2022).

Kugler, M. *Current Developments Current Developments in Immigration Law: The Debate Surrounding "Birth Tourism."* 2018.

Labby, B. Pregnant Calgary Couple "Shocked" after Demand for $10K Upfront Payment to Cover Delivery Costs. *CBC*, Mar. 18, 2022. www.cbc.ca/news/canada/calgary/calgary-immigration-pregnant-uninsured-upfront-payment-1.5173727

Legislation.gov.uk. *British Nationality Act 1981*; Legislation.gov.uk, 2010. www.legislation.gov.uk/ukpga/1981/61

Legislation.gov.uk. *British Nationality and Status of Aliens Act 1914*, Legislation.gov.uk, n.d. www.legislation.gov.uk/ukpga/Geo5/4-5/17/enacted

MacLeod, C. Chinese Flock to USA to Give Birth to U.S. Citizens. *USA Today*, May 1, 2015. www.usatoday.com/story/news/world/2015/04/01/china-usa-birth-tourists-business-strong/24887837/

Mahmudabad, A. K. *The Citizenship Amendment Act—India's Nuremberg Moment*. Times of India Blog, Dec. 21, 2019. timesofindia.indiatimes.com/blogs/counterflows/the-citizenship-amendment-act-indias-nuremberg-moment/ (accessed July 10, 2022).

Ministry of Home Affairs Government of India. *Acquisition of Indian Citizenship*, n.d. Indiancitizenshiponline.nic.in, indiancitizenshiponline.nic.in/acquisition1.htm

Mullen, A. What Was China's One-Child Policy and Why Was It So Controversial? *South China Morning Post*, June 1, 2021. www.scmp.com/economy/china-economy/article/3135510/chinas-one-child-policy-what-was-it-and-what-impact-did-it

National Archives. The Constitution of the United States: A Transcription. *National Archives*, The U.S. National Archives and Records Administration, Sept. 24, 2018. www.archives.gov/founding-docs/constitution-transcript

Neal, C. *The Debate over "Anchor Babies" and Citizenship*; National Public Radio, Aug. 18, 2010. www.npr.org/templates/story/story.php?storyId=129279863

Phillip, A. Inside the Shadowy World of Birth Tourism at "Maternity Hotels." *Washington Post*, Mar. 5, 2015. www.washingtonpost.com/news/post-nation/wp/2015/03/05/the-shadowy-world-of-birth-tourism-at-californias-luxury-maternity-hotels/

PRS Legislative Research. *The Citizenship (Amendment) Act, 2003*; PRS Legislative Research. prsindia.org/files/bills_acts/acts_parliament/2003/the-citizenship-(amendment)-act-2003.pdf

Richardson, C. Op-Ed: A Birth Tourism Crackdown Is Turning U.S. Government Agents into the Pregnancy Patrol. *Los Angeles Times*, Feb. 6, 2020. www.latimes.com/opinion/story/2020-02-06/birth-tourism-crackdown-united-states

RTÉ Archives. *Campaigners for a No Vote in Referendum Disappointed*; RTÉ Archives, n.d. www.rte.ie/archives/exhibitions/1665-immigration/458754-campaigners-for-a-no-vote-in-referendum-disappointed/ (accessed July 10, 2022).

RTÉ Ireland. *RTÉ Ireland's National Television and Radio Broadcaster*; www.rte.ie, Apr. 22, 2004. www.rte.ie/news/2004/0422/49352-citizenship/ (accessed July 12, 2022).

Sajid, I. *South Korea Announces Incentives for More Births*; Aa.com.tr, 2020. www.aa.com.tr/en/asia-pacific/south-korea-announces-incentives-for-more-births/2077452

Stashevska, I. South Florida Sees a Boom in Russian "Birth Tourists." *USA Today*, Mar. 22, 2019. www.usatoday.com/story/travel/destinations/2019/03/22/south-floridas-russian-birth-tourism-boom/3244087002/

US Internal Revenue Service. *U.S. Citizens and Resident Aliens Abroad*; www.irs.gov, n.d. www.irs.gov/individuals/international-taxpayers/us-citizens-and-resident-aliens-abroad

Valdez-Symonds, S. Briefing: Parliament Never Intended for Children to Be Charged Sky-High Fees to Register as Citizens. *Free Movement*, Sept. 4, 2018. freemovement.org.uk/parliament-never-intended-for-children-to-be-charged-sky-high-fees-to-register-as-citizens/ (accessed July 12, 2022).

Wang, F.; et al. The End of China's One-Child Policy. *Brookings*, Mar. 30, 2016. www.brookings.edu/articles/the-end-of-chinas-one-child-policy/

World Bank. *Nigeria Releases New Report on Poverty and Inequality in Country*; World Bank, May 28, 2020. www.worldbank.org/en/programs/lsms/brief/nigeria-releases-new-report-on-poverty-and-inequality-in-country

CHAPTER 11

WHAT'S NEXT FOR DENTAL TOURISM

ABSTRACT

Dental care has been largely overlooked in first-world countries like the United States due to a lack of coverage by insurance providers, insufficient education on the impact oral health has on the rest of the body, and high out-of-pocket pricing, causing widespread rates of tooth decay across the country. However, dental tourism can provide the opportunity for people from countries like the United States to acquire accessible, high-quality dental care for a fraction of the cost. This chapter explores issues regularly affecting consumers across the international dental industry, the socioeconomic factors that can affect oral health upkeep, the impact oral health can have on other parts of the body such as the heart and lungs, and a rundown of some of the top dental tourism destinations globally.

As the fog of the COVID-19 pandemic begins to lift from the travel and medical tourism industry, more and more people across the world are becoming empowered and opening their minds up to new choices and options for receiving dental care. Despite the brief and previously covered pause brought forth to the dental industry by the international health crisis, the dental industry is already back in full swing and likewise is poised for major growth in the years ahead, earning a global $15.57 billion valuation in 2020 that is anticipated to double itself to a $30.59 billion by the year 2027 (Blog.titanwebagency.com, 2022).

With the industry already well on its trajectory toward its forecasted future expansion, as in the cases of most fast-growing sectors, it will likely need to innovate and evolve among growing consumer demand to satisfy said consumer base. According to global consulting firm McKinsey, "when exploring new opportunities, winners go beyond standard focus groups and

Medical, Dental, and Wellness Tourism: A Postpandemic Perspective. Mary Schreiber Swenson, PhD, and Amit Bansal (Authors).

surveys and pull in data on macro trends, marketplace analyses, ground-level performance metrics, and a host of other sources," and "leading companies employ design thinking to create new products and services that will address unmet needs, reach unserved segments, or support entry into adjacent markets" (Cvetanovski et al., n.d.). This will require both relevant private companies and national health systems to adapt to consumer dental trends, or be left behind in the dust for providers who hold satisfying consumer desires as a pivotal business strategy. Dental tourism could become an increasingly popular solution for consumers looking to satisfy their dental needs with quality care at a cost-efficient price.

Before delving into the enduring consumer desires and industry issues that have been affecting the dental sector for decades, it is important to assess how the COVID-19 pandemic itself shifted consumers' perception of wellness, especially under the lens of dental health. A survey conducted on the American public by healthcare marketing agency Lanmark360 revealed that returning to the dentist office for routine appointments was ranked as one of consumers' most important postpandemic visits, surpassing that of visiting the hair salon, the pet groomer, or going out to eat in restaurants (Lanmark360, n.d.). Considering consumers' demonstrated public desire to return to in-person dining, surveyed consumers favoring a future dental visit above eating in their favorite restaurant is incredibly revealing of the public's shifting attitudes toward dental care—especially as, studies show, 6 in every 10 Americans were reportedly scared to visit the dentist's office prepandemic (Deloitte, n.d.; Steingold, 2018).

Furthermore, the Lanmark360 survey revealed that the majority of respondents favor receiving all their necessary dental care in one single visit as of very high importance to them, as well as a majority preferential importance on their dental professionals using the latest and most up-to-date dental technology and equipment. Pandemic-era respondents likewise revealed that, if they did switch dental offices, the most important expectations would be high safety standards, updated office equipment, new technology, better communication, good prices, on-site procedures, modern communication, social distancing, and quality care.

Lanmark360's comprehensive population survey revealed another interesting trend: for the proportion of US respondents who were not planning to book a dental appointment soon after their dental professional's office reopened, while safety concerns were the number one factor behind their reasoning, cost of care was a close second, surpassing respondents who said booking an appointment was not a priority, and those that said they did not need to see a dentist. This is reflective of the US dental industry as whole, which largely runs on out-of-pocket payments—even for the insured.

11.1 THE ISSUES UNDERLYING THE US DENTAL SECTOR

While the US healthcare system is built on the back of private insurance coverage, alongside a variety of public health options for applicable participants including Medicaid and Medicare, dental care is altogether left out of the majority of insurance coverage packages, and these options generally only cover 20% of a patient's dental costs, if that. Instead, dental insurance—much like eye insurance—exists as its own separate entity, and additional packages consumers must purchase if they want their dental visits and procedures to be completely covered by a provider. Considering how deeply intertwined dental health and overall physical wellness are with one another, a topic we will delve into deeper later on in this chapter, dental care theoretically *should* be covered by medical insurance. But in the bureaucratic reality of the US health system, consumers are required to purchase a coverage package exclusively for dental care, or pay for the costs of their treatments and dental visits directly out of pocket.

According to dental sector experts, dental care has likely received its own designated coverage by the US insurance sector due to the fact that dental care is often not perceived as an immediate need and instead as a nonlife-threatening issue that can wait to be reviewed, in contrast to physical medical issues that require an emergency room visit or immediate attention by a physician (Spector, 2017). That is not to say dental problems are not serious—untreated dental conditions can cause numerous lasting problems to the body, especially the longer it takes for them to be addressed. This is reflected in nationwide emergency room trends, which show that there are more than 800,000 annual ER visits brought on by preventable dental issues, which cost an estimated $2.7 billion in 2017 alone (Khazan, 2014; CareQuest Institute for Oral Health, 2021). But in the eyes of the US insurance industry, dental problems are still an issue the consumer must largely address by themselves or through the acquisition of their own dental insurance.

As a result, more than 74 million Americans are left without any form of dental coverage, while the ones that are partially covered by Medicare, Medicare, self-funded insurance, or employer-provided insurance are still forced to cover the large majority of their care themselves. The pandemic provided even roadblocks toward receiving dental care outside of office closures, as a report conducted by the CareQuest Institute for Oral Health revealed that an estimated 6 million US adults lost their dental health coverage throughout the global health crisis, while another 28 million delayed receiving dental care during the same time period due to fears about safety, a lack of insurance, and the high cost of care. Children's dental care is fortunately covered by the Affordable Care Act and is mandatory under state Medicaid coverage, but once they hit legal adulthood, these dental care benefits are stripped away (Pallarito, 2016).

For recipients of Medicare, a plan that covers more than 60 million US residents with long-term disabilities, nearly half of this group receives no dental coverage at all, potentially adding additional health issues to people already suffering from chronic health problems (Freed et al., 2019). The Social Security Act's Section 1862(a)(12) states that "for services in connection with the care, treatment, filling, removal, or replacement of teeth or structures directly supporting teeth, except that payment may be made under Part A in the case of inpatient hospital services in connection with the provision of such dental services if the individual, because of his underlying medical condition and clinical status or because of the severity of the dental procedure, requires hospitalization in connection with the provision of such services" (Social Security Administration, 2018). Medicare makes exceptions for dental services made in relation to another medically classified health issue, such as removing teeth for cancer radiation treatment or making way for the extraction of a tumor, though there are provisions put in place that these procedures must take place at the same time, which ultimately creates an additional burden of stress on the patient to figure out what their coverage includes.

> While the dental exclusion includes exceptions that allow coverage of limited medically-related services, current policy is not completely clear or consistent. For example, Medicare Part A will cover an oral examination for patients who are hospitalized for a comprehensive workup prior to a kidney transplant, but it will not cover an oral examination for patients prior to transplantation of organs other than kidneys. Even in this circumstance, there is no payment under Medicare Part B for the dentist's service, just to the hospital for the service of an employed dentist. As a result of the lack of clarity and consistency in the dental exclusion, there has been ongoing discussion about whether the law can be interpreted to cover a wider range of medically necessary dental services. (Freed et al., 2019)

It is important to acknowledge the average poverty level of Medicare recipients, as socioeconomic factors play deeply into the ability of consumers to pay for their subsequent out-of-pocket dental care costs. According to nonprofit group the Kaiser Family Foundation, 87.9% of Medicare recipients are below the poverty level, with a staggering 37.8% of all recipients reportedly more than 400 times below the established federal poverty level (Kaiser Family Foundation, 2020). Medicare coverage is disproportionately provided to people of color, people with disabilities, rural communities, the elderly population, and other marginalized groups, populations historically overlooked and underserved by the US healthcare system (Gibney, 2020).

Demographical Medicare coverage and the rate of tooth decay across marginalized groups are likely intertwined, as a lack of financial coverage for basic dental care discourages impoverished people from seeking out and paying for that care on their own. As a result, non-Hispanic Blacks and Mexican Americans over the age of 65—groups overwhelmingly and disproportionately represented in Medicare's demographics—are two to three more likely to have untreated cavities or tooth decay than their white counterparts of the same age (Center for Disease Control and Prevention, 2019a).

> Dental insurance may not significantly reduce disparities in access to care if additional barriers to utilization of that insurance exist for African Americans. One such barrier may be that African Americans are disproportionately covered by public (i.e., Medicaid) rather than private insurance. The limited dental services provided by adult Medicaid are associated with a lower likelihood of obtaining dental care and potentially poorer quality care (i.e., a higher prevalence of tooth extractions rather than more expensive root canals) among African American adults. Qualitative studies of caregivers' experiences in attempting to obtain dental care for their children have identified additional barriers to obtaining care, including difficulty finding a dentist who accepts public dental insurance and perceptions that their children experience greater discrimination and poorer quality care because of their public dental insurance. (Schrimshaw et al., 2011)

Regardless of demographic status, Medicaid's adult oral care coverage is slightly more comprehensive than Medicare, and varies on a state-by-state basis, though the criteria for who is eligible for dental care can vary widely based on a number of factors and is not a one-size-fits-all situation. States with no Medicaid dental benefits include Alabama, Delaware, and Tennessee, while 14 states—Arizona, Florida, Georgia, Hawaii, Idaho, Maine, Maryland, Mississippi, Nevada, New Hampshire, Oklahoma, Texas, Utah, and

West Virginia—offer emergency-only coverage to their recipients. Another 17 states, Arkansas, Colorado, Illinois, Indiana, Kansas, Kentucky, Louisiana, Michigan, Minnesota, Missouri, Nebraska, Pennsylvania, South Carolina, South Dakota, Vermont, Virginia, Wyoming, and offer limited dental care under Medicaid, while just 17 provide extensive coverage, including Alaska, California, Connecticut, District of Columbia, Iowa, Massachusetts, Montana, New Jersey, New Mexico, New York, North Carolina, North Dakota, Ohio, Oregon, Rhode Island, Washington, and Wisconsin. While the aforementioned states may have their own form of dental care coverage under Medicaid, that is not to say that all or even the majority of recipients are eligible for Medicaid's oral health coverage, or that the care they receive under said coverage is effectual (Healthinsurance.org, 2022).

An examination of dental industry data reveals that people with preexisting dental insurance coverage are far more likely to attend their routine cleanings and dentist visits, regularly take their children to service their oral health, and overall have better physical health. Two-thirds of those covered by dental insurance, however, are on their employers' healthcare plans, whether through the employers' standard package or through additional benefits bought through the employers' program, which generally do not cover the entire cost of healthcare (University of Illinois Chicago, 2019).

But people who purchase their own dental insurance are not guaranteed to have their dental care covered, either. Current US dental insurance policies generally offer up to $1500 of care per year with annual deductibles between $50 and $100, and "plans typically cover 100% of preventive care, such as regular check-ups and cleaning, 80% of basic services, like filling cavities and pulling teeth, and 50% of extensive work, such as root canals, bridges, and crowns" (Rosato, n.d.). This means two things: a person with good dental health who only needs routine checkups and servicing visits is likely to pay more on their dental insurances' monthly premium than they would paying out of pocket for care, and a person with poor oral hygiene and in need of extensive dental work will most likely have to pay for the majority of their care out of pocket, which can easily climb into the thousands and thousands of dollars range. Though more inexpensive providers exist, the network of dental professionals and offices who take said insurance is limited, which can likewise have a reverberating impact on a patient's overall oral health. Still, as it has been demonstrably shown that having some form of dental insurance motivates people to visit their dental provider obtaining dental insurance can be a net-positive for people otherwise resistant to receiving regular oral healthcare.

Recent trends pulling dentists away from working in their self-owned practices and instead working underneath a consolidated umbrella of providers has only shot up the price of dental insurance and made it even more unattainable for many members of the US population.

> A dentist's income is tied to the number of patients treated and the reimbursement rate received for those treatments. However, higher reimbursement rates may pressure dental insurers to charge higher insurance premiums in order to maintain the same level of profitability. If employers, who pay for most of the commercially insured dental insurance plans, find these premiums to be too high, they could switch to another dental insurer or drop dental coverage altogether. This means that dental insurers have an incentive to put downward pressure on reimbursement rates whereas dentists aim to receive more compensation for their services. The net effect on provider prices depends on the relative bargaining strength of providers and insurers. In medical markets, concentrated insurance markets are also associated with larger physician practices and a greater likelihood that physicians are directly employed by hospitals. As a result, consolidation of providers has counteracted potentially lower prices that can arise from health insurance consolidation. The net result is higher prices for medical services. (Nasseh et al., 2019)

While we have previously discussed whether healthcare is a right or a commodity in the United States, when it comes to oral healthcare, one thing is clear: in America, dental care is a commodity, and one many of the population simply cannot afford. Considering the lasting effects consistent dental care and hygiene has on a person's overall physical health, America's broken dental health system and widespread prominence of tooth decay could potentially be the hidden culprit behind some of the US population's enduring health issues.

11.2 HOW DENTAL CARE AFFECTS OVERALL HEALTH

Oral health status can have lasting impacts on a patient's overall well-being and has been established by the Office of Disease Prevention and Health Promotion as one of 10 leading health indicators that needs to be addressed in the next decade to improve the health of the US population (Office of Disease Prevention and Health Promotion, 2014).

Given the mouth's vital role as a vehicle for eating, breathing, and talking, the mouth is perceived as a "gateway" to the rest of the human body, where not only food and drink can enter, but plentiful bacteria as well (Dillenberg, 2019). When filled with good bacteria, the mouth's natural microbiome flourishes and assists with the upkeep of dental health. However, as the mouth has a direct entrance to both digestive and respiratory systems, bad bacteria can sometimes creep in and infect the body with disease. Likewise, the mouth's vascular quality and incredible internal network of blood vessels can contribute to the spread of disease, as bacteria's entrance into the bloodstream is only separated by a very thin sheet of tissue.

While consistent dental hygiene and care, like brushing and flossing, can help stave off this bacteria and uplift oral health, failing to maintain a regular oral hygiene regimen can create a breeding ground for bacteria both in the mouth and other parts of the body and ultimately lead to or exacerbate disease.

According to dental and medical experts, the following medical conditions can be exacerbated or contributed to poor dental health, or, on the flip slide, go on to negatively affect a patient's dental health:

Cardiovascular disease: Heart disease is already a prominent issue across the US healthcare system, ranking as the number one cause of death for American adults and costing the American Healthcare system approximately $229 billion annually. According to recent statistics from the Center for Disease Control, one person dies every 34 s in the United States from heart disease, turning cardiovascular disease into a crisis in its own right (Center for Disease Control and Prevention, 2019c). Poor dental health is viewed as a major contributing factor toward the development of heart disease, when the heart's valves and chambers are infected by mouth-born bacteria spreading through the bloodstream and eventually attaching in the heart in the case of endocarditis. Gum disease—also known as periodontitis—is likewise associated with an increased risk of cardiovascular problems, while tooth loss has been found to correlate with coronary artery disease (Salinas, n.d.). Research from the Cleveland Clinic states that "some studies have shown that bacteria in the mouth that are involved in the development of periodontal disease can move into the bloodstream and cause an elevation in C-reactive protein, a marker for inflammation in the blood vessels," which can lead to cardiovascular problems and strokes, while maintaining a consistent dental hygiene regimen can at least prevent additional contribution to the development of heart disease (Cleveland Clinic, n.d.). Dental tartar and calculus is also an easy way to spot risk of a heart attack, revealing a build-up of calcium that is likewise impacting other parts of the body, like the heart, kidneys, and prostate (Dillenberg, 2019).

Pneumonia: Mouth-born bacteria and mouth disease can also contribute to the development of respiratory diseases and infections, with research studies demonstrating the link between poor dental health hygiene and pneumonia, one of the global population's top killers. "One cubic millimeter of dental plaque contains about 100 million bacteria and may serve as a persistent reservoir for potential pathogens, both oral and respiratory bacteria. It is likely that oral and respiratory bacteria in the dental plaque are shed into the saliva and are then aspirated into the lower respiratory tract and the lungs to cause infection" (Paju and Scannapieco, 2007).

Diabetes: People suffering from chronic diabetes—or more than 9.3% of the American population—may be at heightened risk for poor dental health due to the condition's symptoms. Diabetics may suffer from a lack of saliva production and make the mouth feel dry; as saliva is an important protector of the teeth, a lack of sufficient saliva can set the stage for increased tooth decay among the diabetic population. Diabetes similarly comes with an increased likelihood of gun-related dental issues like gingivitis and periodontitis, a chronic, inflammatory disease that, if untreated, can eventually destroy the teeth and the surrounding gum tissue needed to keep them in place. In fact, some 22% of all diabetics are suffering from periodontitis, according to the American Dental Association. Gum disease then has the potential to exacerbate diabetes symptoms in turn, making blood sugar levels more difficult to control and subsequently making diabetics more susceptible to infections (Martin, 2019).

Pregnancy: Dental care becomes of heightened importance for pregnant women, whose dental hygiene regimen can have impacts on a pregnancy's ease and the postbirth health of the child. Pregnant women are more susceptible to contracting gum disease or dental issues due to the hormonal changes experienced throughout the course of fetal development, leading to the common formation of "pregnancy tumors"—which are essentially lesions—on the gums. Pregnancy tumors generally form in the mouth around the second month of fetal development, reach their peak inflammation in the eighth month, and generally go away shortly after the birthing process has been completed, though they may need to be surgically removed if they do not organically disappear. Approximately 40% of pregnant women experience some form of gingivitis throughout fetal development, though pregnant women with well-kept dental hygiene are less likely to contract gingivitis. Tooth decay also increases throughout pregnancy, especially considering the customary nausea and vomiting spells that many women face in the first trimester, as the stomach acid brought up from the stomach begins to erode away at tooth enamel. Research has found that if bad oral hygiene is in place, "pregnant women can experience premature delivery, low birth weight baby, pre-eclampsia, gingival tissue ulcerations, pregnancy granuloma, gingivitis, pregnancy tumors (epulis gravidarum), loose teeth, mouth dryness, and dental erosions" (Yenen and Ataçağ, 2019).

HIV and AIDS: People with HIV and AIDS are disproportionately affected by dental disease, as autoimmune disorders make it difficult for the mouth to naturally retain a health microbiome. HIV and AIDS attack the body's T cells, which are vital to maintaining a functioning immune system, and subsequently, good dental health. The mouth is actually one of the first places HIV and AIDS lesions begin to form, and these are important signifiers to health professionals that a patient is carrying the disease. As a result of a weakened immune system, carriers of HIV and AIDS are susceptible to mouth-related issues like ulcerative periodontitis, Karposi's Sarcoma, red band gingivitis, oral hairy leukoplakia, dry mouth, thrush, canker sores, mouth sores, and herpes outbreaks (American Dental Association, n.d.). Oral cancers are also prevalent across the United States, with 54,000 new cases of oral cancer diagnosed in 2021, and 11,230 deaths caused by the disease during the same year (American Cancer Society, 2019).

Cancer and tumors: There have been numerous reports of the link between gum disease and cancer across recent years, with a decade-long study published in the *Journal of the National Cancer Institute* revealing that there was a 24% increased risk for developing cancer for patients with severe periodontitis, as opposed to those with mild or no form of the gum disease (Johns Hopkins Medicine, 2018). Likewise, traces of periodontal bacteria have been found in the biopsies of cancerous tumors, including *Treponema denticola* in pancreatic tumors, *Fusobacterium nucleatum* in colon cancer, and *Porphyromonas gingivalis* in oral cavity tumors (Dillenberg, 2019).

Oral health clearly evidently has a far-reaching impact on the human body beyond the scope of the mouth, and dental and gum disease certainly is not limited to the groups with the conditions above. Statistics from the National Institute of Dental and Craniofacial Research show that more than 92% of US adults between the age of 20 and 64 have had some form of dental cavity in their permanent teeth throughout their lives, whether treated or untreated, while more than one in every four adults in this age group—26%—have untreated dental caries (National Institute of Dental and Craniofacial Research, 2018). Similarly, more than 47.2% of adults over the age of 30 have some form of periodontal disease, such as periodontitis or gingivitis, with the statistics only growing more likely with age (Center for Disease Control and Prevention, 2019b).

But with the general population largely uninformed of the lasting consequences of their poor dental care, dental health is not treated as the vital practice it *should* be in the United States. It is difficult to motivate the population to take their dental health seriously given the US dental sector's numerous financial roadblocks and lack of insurance coverage. If dental health is to be taken seriously, there need to be readily available solutions in place to provide affordable and attainable access to dental care for the people who need it most.

11.3 ADDRESSING THE MODERN DENTAL SECTOR THROUGH MEDICAL TOURISM

Medical tourism provides an intriguing opportunity for consumers from the United States to find quality dental care at a price they can afford out-of-pocket, shirking the need for dental insurance. Dental tourism is not unique to the United States, however; the phenomenon is demonstrable across the world and has roots all across Central and South America, Asia, Europe, and Africa—not to mention the United States' northernmost neighbor, Canada.

Dental tourism allows patients of all backgrounds to take power back into their own hands by crossing borders, allowing consumers to stand up against the conventionally available options in how they choose their dental care options and control cost. As a result, the dental tourism industry is beginning to take off and rapidly expand and is estimated to reach a $24.3 billion valuation by 2032, more than quadrupling dental tourism's projected value of $6.7 billion in 2022 (Future Market Insights, 2022).

As previously explored, one of the main factors discouraging US citizens from seeking out dental care is cost. Though travel costs may seem intimidating to medical tourism novices, dental tourism can slash the costs of necessary treatments by up to 80% depending on the destination, most likely covering the cost of travel while still providing additional savings straight into consumer's pockets. Plus, dental tourism often slashes waiting times that have become standard in the United States and expands consumers' provider options to a much wider pool.

More Americans participate in the dental tourism industry than in the traditional medical tourism sector, says data from Healthcare.com, which reveals that 470,000 Americans went abroad for the purpose of receiving dental care in 2019, compared to 310,000 who traveled for medical purposes. Despite the arrival of the COVID-19 pandemic in 2020, both dental tourism and medical tourism still forged ahead, albeit at a scaled back rate, and US dental tourists once again outpaced medical tourists with registered 170,000 dental tourists, compared to 130,000 medical travelers (Rivera, 2021).

US dental patients benefit from the nation's close proximity to Mexico, one of the leading destinations for dental tourism around the world. Accessible by car or by plane, Mexico's close proximity to the United States and standard six-month visa waiver for American citizens has encouraged hundreds of thousands of US tourists to cross the border into Mexico every year to receive dental care (Da Silva, 2019). The border town of Los Algodones, affectionately monikered as "Molar City," draws in a multitude of American dental patients every year; with 350 dentists located in the tiny town, Los Algodones has the highest number of dentists per capita in the entire world and has become an easy to access destination for Americans looking for affordable dental care only a few mile drive across the US–Mexico border (Jacobs, 2018).

Mexico's open-border policy to Americans makes it impossible to ascertain the true amount of US adults who head to the country for dental services, leaving many cases of potential dental tourism unregistered for statistical analysis, though some industry experts say the number reaches as high as more than 1 million US dental tourists per year.

High-quality dentistry is able to be cheaper in Mexico than the United States due to the Mexican government providing numerous subsidies to finance Mexican dentists' education. Without the need to pay off student loan debt, which is a recurring theme throughout the world of US higher education and was an estimated $301,583 per dental school graduate in 2021, Mexican dentists do not need to charge exorbitant fees to their clients to recoup educational costs (American Dental Education Association, n.d.; Robbins, 2014).

As a result, Mexican dentists are able to charge service fees much lower than that of their US counterparts; in Mexico, single-tooth dental implants start at $700, compared to $1500 in the United States; all-on-4 implants start at $6000, compared to $20,000; acrylic dentures at $350, compared to $1000; dental bridges at $600, compared to $1500; and root canals at $300, compared to $700 (Dentaly.org, n.d.). As private providers, US dentists can go on to charge far more than the aforementioned starting cost of treatment, and true cost varies from provider to provider. Regardless, the cost savings of receiving treatment in Mexico are demonstrably favorable to consumers, especially for large-scale work like all-on-4 implants, and can be a simple solution for Americans without dental insurance or a lack of available funds to pay out of pocket for US dental care.

Mexico is not the only available and popular dental tourism destination for Americans and other foreign travelers, with other countries providing high-quality and cost-effective dental care including Panama, the Dominican Republic, Costa Rica, Thailand, Hungary, and India, providing numerous options for dental tourists spanning the globe (Live and Invest Overseas, 2022). Domestic dental tourism has also begun to take off, offering opportunities to receive better quality and cheaper care from providers located in a country's different states or provinces.

Still, consumer misconceptions about dental tourism and travel for treatment have instilled a fear in the American population from participating in the sector and seizing the fate of their dental care into their own hands. A recent survey from RetireGuide published in July 2022 revealed that three out of four US adult respondents between the ages of 18 and 65 fear having dental work performed in a country outside of the United States, with only 10% of this group saying they would push past their fears to receive treatment abroad (Hanson, 2022). Another 24% of Americans revealed they would participate in dental tourism if they were treated with modern medicine, putting the US public's misunderstanding of international health standards on display, as most popular dental tourism destinations are equipped with dental professionals and offices on par with US standards and many countries, including Mexico, have multiple destinations with Joint Commission International accreditation.

Take a study on the Hungarian dental tourism industry, for example, which revealed the demonstrable efficacy of the popular dental tourism destination's providers.

> There are several reasons why Hungary could maintain a leading position in dental tourism. First, the cost/benefit ratio is outstanding. The affordable price and value for money were already recognized in the early 1990s and were appreciated even before Hungary joined the EU. Second, the high quality of Hungary's dental profession: (a) dental professionals in Hungary are well-qualified dental practitioners who have received high-level dental training; (b) dental professional standards are up-to-date and often supervised; (c) in Hungary dental technology, the quality of materials, and equipment used in dental practices is of European level quality. The rate of complications in dental care is around 5%, similar to other European countries. Finally, previous treatment experiences are positive and patient satisfaction levels are high. More and more patients seek care in Hungary, and more and more patients state that they would be willing to return for further treatment in the future. (Kovacs and Szocska, 2013)

There needs to be a massive reshaping of the American public's perception of dental tourism in order for the practice to become normalized and to reduce the population's fear of dental tourism. Concerns about safety and the cost of travel are not completely unwarranted, but can easily be aided by an experienced liaison in the dental tourism industry, who can set up patients with their best-fit destination and introduce them to reputable providers that have been vetted for the quality and standard of their care.

One up-and-coming online platform, Dr. Mary Swenson's My Smile Choices, looks to make finding dental care for patients more accessible and easier than ever through the My Smile Choice's streamlined search interface. Here, patients can browse through vetted providers across the United States and Central America, shop around for the right dental professional for their unique cases, and find easy access to top dentists across the world at the click of a button—even on holiday. Offering a simple, experienced, and easy-to-navigate solution for acquiring dental care, platforms like My Smile Choices could be the key next step in democratizing dental care.

The US dental healthcare system is evidently in need of a large overhaul if it intends to take care of its citizens' healthcare—but that does not mean consumers need to wait around for care in the meantime. Embracing dental tourism gives the power of choice back to patients, offers massive cost savings, and still provides high-quality results. As the medical industry continues to evolve in the postpandemic reality and consumers begin to understand the inextricable link between dental health and overall well-being, expect clear-cut solutions like dental tourism to continue an upward growth trajectory in the years to come.

11.4 THE WORLD'S TOP DENTAL TOURISM DESTINATIONS

Bulgaria: Bulgaria has become a major destination for dental tourism over recent years, offering up to 50% discounts on standard dental care on procedures compared to pricing across the rest of the European continent. The educational standards for Bulgarian dentists are high, requiring dental professionals to complete six years of school before they are eligible to practice dentistry. Because Bulgaria is located within the European Union, Bulgarian dentistry is held to the same high standards as the rest of the continent, and all Bulgarian dentists and practices must be audited on an annual basis in order to continue their practice (Dentaround, n.d.).

International dental tourists have become enticed by Bulgaria's multilingual translation options; due to its geographical locations, many Bulgarian dental professionals are fluent English speakers, while Russian and German are also commonly accommodated in the Bulgarian dental sector. Bulgaria likewise has a robust ratio of dentists to its population, boasting more than 100 dentists per 10,000 inhabitants—one of the top five highest concentrations of dentists in the world (World Health Organization, 2022). Because of this high number of dentists, the Bulgarian dental industry is able to accommodate new patients and dental tourists with little to no waiting time (Impey, 2011). With numerous low-cost European airlines offering direct routes to Bulgaria, plus Bulgaria's wealth of natural attractions, historical sites, and gorgeous beaches, Bulgaria has become an ideal destination for dental tourists from across the world looking to receive high-level European dental care at a fraction of the price, with a vacation to boot.

Pricing: In Bulgaria, titanium implants reportedly begin at $600, compared to $3500 in the United States; all-on-4s at $3145, compared to $24,000; and root canals at $65, compared to $700 (Goel, n.d.).

Cambodia: Dental tourism has become a targeted sector for growth by the Cambodian government, prompting the country to welcome dental tourists from across the world into its borders with open arms. Reports from 2015 show that 90% of Cambodian private practice dental revenue was from international tourists, a number only likely to have increased over recent years as Cambodia continues to evolve and embrace dental tourists. Cambodia's dental professionals and practices are regulated by the Cambodia Dental Association, an independent nonprofit boasting membership from more than 170 members of the Cambodian dental community (Cooperation Committee for Cambodia, n.d.). While Cambodia's medical industry is still in its fledgling stages, many of its dental professionals head to nearby countries like Thailand to receive their dental education, and while major progress is being made in its forward development, Cambodia's medical industry is still shy from meeting the standards of other popular dental tourism destinations (Bite Magazine, 2013). This, in turn, can allow for cheaper costs for dental patients; because Cambodian dentists are not required to obtain licensing for x-ray machines, insurance for potential malpractice or litigation, and no need to register with a dental board, plus Cambodia's significantly lower overhead costs and cost of living, prices for common dental care and procedures can be slashed by as much by 80% when compared to the United States. The low cost of living in Cambodia has reportedly attracted numerous Western dental professionals to practice in the country, including those from the United States and Japan (Dental Travel Services, n.d.).

Despite potential concerns about the quality of care, which is purportedly improving year over year, Cambodia's star in the international dental tourism sector is on the rise: Cambodia's close relationship with Western counties in its modern development has made English speakers bountiful throughout the kingdom, particularly among specialists like dental professionals, attracting English-speaking tourists into the country to receive care. Unbeknownst to many, Cambodia actually has two official currencies, the Cambodian Riels and the US dollar, making transactions particularly breezy for American tourists (United Kingdom Government, n.d.). Cambodia is accessible via air or by land, with many tourists electing to take buses or taxis from the neighboring countries of Thailand, Laos, and Vietnam. Cambodia is similarly filled with a rich cultural history that may attract dental tourists to extend their stays, including Angkor Wat, which is widely considered one of the wonders of the world, and its famous beautiful beaches.

Pricing: Due to the aforementioned factors, dental care in Cambodia is incredibly inexpensive in comparison to the rest of the world. According to one of Cambodia's leading dental practices, crowns start at $150, root canals at $50, and fillings at $10, prices up to 80% cheaper than offered in the United States (Pagna Dental, n.d.).

Colombia: Colombia has gained global renown as one of the top destinations for dental tourism in the world, bringing in high-profile customers like American rapper Rick Ross to acquire dental work in the country, and subsequently gaining major attention for its dental care offerings in the international media (Tigg, 2020). Colombia's dental tourism sector pulled in more than $140 million in 2019 alone and is projected to bring in $226 million by the year 2024 as more and more tourists elect to travel to Colombia for dental care (Mendoza, n.d.).

Dental care prices are around 70–80% lower than US costs on average, making Colombia a cost-efficient selection for many international tourists. This is possible due to numerous factors; Colombian dental professionals generally graduate from their education with no student debt, allowing them to charge lower prices than US dentists that are saddled with hundreds of thousands of dollars' worth of student debt; overhead costs are low, including rentals, services, and salaries, allowing dental practices to build out an extensive team of highly trained professionals at a marginally lower cost than the

United States; and the cost of living is low, creating lower healthcare and dental prices in turn.

Though dental care in Colombia may be affordable, that does not mean Colombian professionals are not highly qualified, as all Colombian dentists are required to go through a mandatory 5-year study to earn their dental diploma and practice dentistry professionally, followed by 1 year of public service, and there are more than 21 accredited dental schools in Colombia as of 2014 (Jaramillo et al., 2010). Many Colombian dentists offer warranties for their procedures, which can span from multi-year to lifelong depending on the type of procedure, giving international dental tourists peace of mind that their results are covered in the event of an accident in the years post-treatment (Jacobs, 2021). Likewise, the Colombian healthcare system as a whole is highly rated, and earned the number 22 spot out of 191 on the World Health Organization's international healthcare system rankings, beating out both Canada, ranked at 30, and the United States, ranked at 37 (Doyle, n.d.).

Thanks to its close proximity to the United States and position at the northeastern most part of South America, Colombia has become an easily accessible and central destination for patients from all across the Western Hemisphere, and many Colombian dental professionals are proficient in English as well as fluent in their native Spanish. Colombia's historic cities, local cuisine, beautiful coastlines, and abundant flora and fauna have likewise poised the country as a top tourism destination and have led many dental tourists to expand their travel for treatment to Colombia to encompass sightseeing and vacationing.

Pricing: All-in-4 dental implants cost around $5500 in Colombia, compared to $25,000 in the United States, providing savings of 78%. Snap-on dentures cost $2100, compared to $12,300 in the United States, while root canals can be obtained for as low as $110, compared to the United States starting price of $700 (Dental tourism in Colombia, n.d.).

Costa Rica: Costa Rica's expansive dental sector has turned the Central American country into one of the world's most sought-after dental tourism destinations, with a recent report from EuroMonitor revealing that 90% of the nation's dental procedures in Costa Rican hospitals are performed on international tourists (TourismReview.org, 2019). The large influx of international tourists into Costa Rica is likely in part due to the country's close proximity to the United States and Canada and is easily accessible from both via a 4–7-h flight. Costa Rica's consistently low crime rate and stunning natural attractions have also impacted the country's international appeal and impacted international tourists' decisions to receive care in the Central American nation (Madigan, 2019).

The Costa Rican dental industry is highly modernized and on par in terms of technology and equipment with that of the United States, with Costa Rican dental clinics purportedly importing equipment from leading US providers like Bicon, Keystone Dental, Straumann, and Biocare (Dental Implant Cost Guide, n.d.). All Costa Rican dentists go through a five-year educational process and must acquire licensing from the National College of Dental Surgeons before they are able to legally practice dentistry in the country. Many Costa Rican dentists are educated outside of Costa Rica and in the United States or other countries with extensive medical educational opportunities, and Costa Rican dental professionals are known for being multilingual—a major draw for international patients looking to receive care (Lopez, n.d.). Like Colombia, many Costa Rican dental practitioners also offer life time or multiyear warranties on their care, standing in stark contrast to the methodology of the US dental sector and have likewise become known for offering transparent pricing models to its patients, allowing for Costa Rican dental care to cost significantly less than in the United States (Alvarado, n.d.).

Pricing: All-on-4 implants cost $11,500 on average in Costa Rica, more than half the price of the procedure in the United States. Full porcelain crowns cost $500 on average, while porcelain veneers have a price tag of $500 per tooth—$1250 less than the US average of $1850 (Goodman, 2020).

Czech Republic: As Eastern Europe's medical tourism industry continues to expand at a 12–15% annual rate, the Czech Republic has become one of the region's standout destinations for obtaining high-quality dental care (Hamper, 2018). Located in Europe's Schengen area, the Czech Republic's immigration policies allow US citizens to travel to the country without needing to apply for a visa, while the Czech Republic is similarly easily accessible for Europeans by air or by land. The Ministry of Health of the Czech Republic has built a searchable database of all registered medical and dental professionals in the country, allowing for patients to shop around and choose from available providers to find the best care possible to fit their unique cases (Karolyi et al., 2019). Traveling to the Czech Republic under the lens of medical tourism also opens up the opportunity for patients to turn the trip into a holiday: the country boasts 12 UNESCO Heritage Sites within a multi-hour drive from Prague, plus numerous historic sites to experience while within the capital itself.

Dental professionals are held to high standards in the Czech Republic and must complete two years of postgraduate training after completing their educational requirements, after which they must pass an exam in order to be officially licensed to practice. All dentists are required to join the Czech Dental Chamber, helping ensure the continued quality and efficacy of their work and providing additional restaurants for international dental tourists concerned about obtaining dental work in the country (CzechTourism, n.d.). The Czech dental industry has become reputed for its use of modern techniques, state-of-the-art technology such as CEramicREConstruction, and advanced equipment on par with higher-priced countries like the United States, and its dental professionals are known for being proficient in numerous languages to accommodate foreign patients, including English.

Pricing: Dental work is more expensive in the Czech Republic capital of Prague than in other parts of the nation, while still marginally less expensive than in the dental systems in the United States and the United Kingdom, and dental care goes on to cost less as patients travel further from the capital. The average cost of a crown is \$500, \$100 for a root canal, and \$890 for dental implants, slashing the costs of US dental care by more than half (Dental Tourism Expert, n.d.).

Germany: Germany has become a trusted source of dental care for dental tourists from across the world thanks to its strict regulatory standards, high quality of care, and slightly lower costs when compared to markets like the United States and Western Europe, though pricing in Germany is still higher than in Eastern European countries like Hungary and the Czech Republic. Still, Germany's renowned dental system has made it a major destination for nearby European countries where dental care is considerably more expensive, particularly Switzerland (Gheorghe et al., 2017). Americans can likewise travel to Germany with ease to receive dental care thanks to the country's location in the Schengen area, and the country's rich history and culture have made it an attractive tourism destination beyond the scope of dental care.

Germany's National Association of the Statutory Health Insurance Dentists gives patients a list of accredited dentists across the country, helping dental tourists gain a better scope of choice in what provider they would like to visit and what specialties each professional performs (Association of the Statutory Health Insurance Dentists, n.d.). The country's wealth of acclaimed dental schools has created a population of high-trained dental professionals throughout Germany, all of which must go through years of education followed by multiple examinations before they can practice dentistry professionally (Edupark, n.d.). Most German dental professionals speak English and can accommodate international clients accordingly.

Pricing: Dental costs in Germany have become increasingly cheaper to US consumers throughout 2022, as geopolitical events have brought the Euro and US dollar to near equivalent trading value. As a result, dental implants start at $1800, root canals at $450, and crowns at $600, pricing slightly lower than offered in the United States and competitive in comparison to the dental markets in nearby countries like Switzerland (Honova, 2022).

Hungary: Hungary is tied with India for the world's second most popular destination for dental tourism behind Mexico, pulling in patients from across the world with its signature cost-efficient care and high-quality treatments for more than two decades (Peručić, 2019). Another member of the Schengen area, American patients can easily travel to Hungary without a visa to receive dental care, while Hungary's borders are similarly open to all members of the European Union. Hungary has become known for its use of technologically advanced equipment, treatment materials up to the standard of other European countries, its emphasis on customer care, excellent facilities, and numerous sightseeing destinations for postcare tourism.

The demographics of dental tourism patients varies depending on what region of Hungary is being visited; according to research studies, countries near Hungary tend to dominate dental tourism in the country's western region, while dental tourism in the Hungarian capital of Budapest consists mainly of international patients from countries from much farther away.

> ….the provenance of foreign dental care patients shows a much more heterogeneous distribution [in Budapest] compared to Western Hungary. About one fifth of foreign patients treated in Budapest are from the UK followed by Austrians (17.4%), Germans (11.8%) and Swiss customers (11.8%). In relative terms, 97.5% of UK dental patients in the target regions are treated in Budapest. The respective figure is 37.9% for Germany and 30.2% for Switzerland. In the Austrian case, just 6.2% are treated in Budapest. (Österle et al., 2009)

Receiving dental care in Hungary can open up a world of procedure-specific tax breaks for other Europeans who travel to the country under dental tourism, benefits that aid citizens of France, Ireland, and numerous other countries across the continent, helping propel Hungary into Europe's number one destination for dental tourism (Medical Tourism Review, 2020).

Hungary's dental market has expanded in response to this increase in demand, with the number of dentists in the country increasing by 56% between the years 2000 and 2008 (European Union Cross Border Care Collaboration, 2013). All of Hungary's leading medical universities offer dental programs, resulting in a population of proficiently trained dental professionals; in fact, the quality of dental education in Hungary is so widely regarded that there is an outflow of Hungarian-trained dentists into other countries across the world. Dental professionals are regulated by Hungary's National Advisory Board of Healthcare, helping ensure Hungarian dental professionals maintain high standards of care throughout their practices. As a result, Hungary has high rates of patient satisfaction, and a complication rate of 5%, which is on par with other European countries (Österle et al., 2009).

Pricing: The cost of dental care is relatively cheap in Hungary, though the cost of travel to the country likely makes Hungary a more apt destination for nearby Europeans, who can arrive by train or budget airfare, than Americans who need to take a trans-Atlantic flight. In the Hungarian dental market, single-tooth implants are priced at $600 on average, porcelain veneers at $300 per tooth, and full-mouth implants beginning at $2300, prices 70% cheaper than the standard costs in the United States (Asmussen, 2022).

India: Joining Hungary as the second most popular dental tourism market in the world is India, which offers affordable costs to medical tourists up to 90% lower than those offered in the United States and the United Kingdom. With a robust English-speaking population, India's cost-efficient dental care and linguistic accommodations has attracted international patients from across the globe. Likewise, the Indian government has created an "M class" tourist visa for international visitors looking to enter the country for the explicit purpose of receiving medical or dental treatments, making the immigration process into the country remarkably streamlined (Ruddh, 2018). India's incredible wealth of sightseeing opportunities, historic landmarks, and culture has similarly enticed travelers to extend their dental tourism into planning a more expansive trip, bolstering India's tourism sector in turn.

India has multiple provisions in place to ensure the quality of its dental professionals, requiring all professional dentists to achieve a Bachelor's in Dental Surgery and complete a year of internship before receiving accreditation from the Indian Dental Association, the country's independent body of dental professionals (Indian Dental Association, n.d.). Many dental professionals in India study outside of the country to receive their education in Europe or the United States, creating a global perspective on dentistry industry-wide in India and providing international patients with care up to the standard of their country of residence. The adoption of state-of-the-art technology and equipment has helped propel the Indian dental sector toward continued growth, while also providing an additional attraction to dental tourists (Vijay, 2019).

Pricing: In India, crowns start at $70, veneers at $110 per tooth, root canals at $90, and $440 per tooth for dental implants, offering savings up to 90% of what patients can receive in the US dental market (Napitu, 2022d).

Indonesia: A more recent addition to the international dental tourism industry, Indonesia is beginning to become a major destination for dental tourists across the world, especially as more and more consumers are becoming aware of Indonesia's beautiful attractions, sightseeing opportunities, and robust selection of high-end resorts and hotels. As a result, Indonesia has become a simple choice for patients looking to couple their high-quality and cost-efficient medical care with an extended vacation—particularly in Bali and Jakarta.

Indonesia has especially become a germane choice for Australian dental tourists due to its close proximity and reasonable flight time; according to the president of the Indonesia Dental Association, 45% of all Australian dental tourists chose Bali as their preferred destination for dental treatment abroad (Akbar et al., 2020). With more than 21 different airports to fly into across the country, Indonesia offers a multitude of tourism options and dental providers no matter where its incoming international dental patients hail from, and Indonesia's 30-day visa waiver policy for 169 countries eliminates tricky immigration barriers from the equation (Indonesia Secretariat of Tourism, n.d.). Many Indonesian dental providers offer their clients full scope packages covering transportation, accommodation, procedures, after-care, and tourism additions, simplifying the planning process for tourists looking for more streamlined dental tourism options.

While dental standards for the general Indonesian population are unfortunately low and filled with self-taught dentists, the quality of care in the private sector—which generally covers Indonesian dental tourism—is markedly better (Yuniar, 2021). Still, as in all cases of dental tourism, it is important to carefully evaluate the available providers and thoroughly vet options before booking a dental tourism trip to Indonesia.

Many high-end practices in Indonesia offer the same full-scope of services provided by dental practices in Europe, Australia, and the United States, including Invisalign, which are accompanied by modern technologies and equipment. But given the low cost of living and overheads in Indonesia, common dental procedures can be obtained by dental tourists in Indonesia for a fraction of the cost of what it would be in their native countries (www.expat.or.id, n.d.).

Pricing: In Indonesia, common procedures like crown and bridges, root canals, and porcelain veneers begin on average at $284, $198, and $379 per tooth, nearly one-tenth of the price these procedures cost in Australia, Indonesia's largest source of dental tourists (Aesthetics Dental Care, 2020).

Mexico: Thanks to its close proximity to the United States, Mexico has propelled into the globe's number one dental tourism destination worldwide. According to recent projects, the Mexican dental tourism sector is continuing to expand and is forecasted to reach a $45 million valuation by the year 2027 for just implants alone (iData Research, 2021). Mexico is just a quick flight or drive across the border for most US dental patients, where the majority of Mexico's dental tourists source from, while the country's bustling tourism sector and multitude of airports have made Mexico's dental industry accessible for patients hailing from all corners of the map. With plentiful beaches, a rich and history-filled culture, and a bustling culinary scene, Mexico's bountiful tourism attractions only stimulate the destination's popularity for dental tourism.

Mexican dentists are required to go through three to five years of dental education and one year of community service before they are able to practice professionally, and, as of 2017, there were 156,622 licensed dentists in Mexico, 99% of which serve the private sector (Geib, n.d.). Prices are kept down by the low cost of living and low overhead costs, while the quality of care remains on par to that of the United States.

Pricing: In Mexico, single-tooth dental implants start at $700, all-on-4 implants start at $6000, acrylic dentures at $350, and root canals $300, compared to $1500, $20,000, $1000, and $700, respectively (Dentaly.org, n.d.).

Philippines: The Philippines has become a recent player in international dental tourism, much of which can be attributed to its ex-pats, many of which return to their native Philippines to receive cost-efficient and quality dental care at a lower cost than they can obtain in their present countries of residence. The country particularly appeals to dental tourists hailing from New Zealand and Australia due to proximity but extends to the rest of international consumers considering the potential savings of up to 75% the Philippines' dental industry can provide. Though its location in the North Pacific may not be convenient for most outside of the region and Oceania to travel to for the specific purpose of dental tourism, the Philippines' tropical climate, 7600 islands, and numerous tourism attractions make the destination a perfect fit for dental care for people who already have plans to travel there, or are looking for a two-for-one trip encompassing both vacationing and medical care (Napitu, 2022c).

Dental education in the Philippines is designed to produce high-quality and properly trained dental professionals, allowing many providers in the Philippines to offer care that meets the standards of incoming international dental tourists, though a robust vetting process of providers should still be followed for anyone planning a dental tourism trip to the Philippines.

> The dental schools of the Philippines are regulated by the Commission on Higher Education (CHED) in the Philippines. Moreover, CHED Memorandum Order (CMO) No. 33 Series of 2006 stipulates the policies, standards, and guidelines for dental education. Universities that anticipate the offering of the 6-year DMD degree program must adhere to the former and make the latter a guide. This program is composed of the predental course in the first two years, basic medical and dental sciences with pre-clinical subjects in the next two years, and the clinical training in the last two years. (Dalanon and Matsuka, 2019)

Pricing: In the Philippines, all-on-4 implants start at $8000 on average, single-tooth implants at $1000, and porcelain veneers at $260 per tooth, some 75% less expensive on average than the equivalent procedures performed in the United States.

Poland: Poland is another one of Eastern Europe's most popular medical tourism destinations, and reportedly brought in more than 100,000 international tourists for the express purpose of receiving dental care in 2012 alone (Brignall, 2014). The Polish government has made a concentrated effort into becoming one of the world's top medical tourism hubs, which has subsequently enhanced the country's dental tourism sector in turn. As part of the European Union, Polish dentists are held to strict quality standards and most Polish dental facilities are modernized, clean, and equipped with the latest dental technology. Its Schengen area designation lets American patients enter the country with ease and is easily accessible through its major international airport in Warsaw, while Poland has likewise become a sought-after destination for European tourists hailing from the United Kingdom, Germany, Switzerland, and Belgium.

While Poland's public dental sector can experience lengthy waiting times for Polish citizens, the country's private dental sector can offer little-to-no wait time for procedures for international tourists and Poles interested in participating in the private sector. Holding customer service in high regard, many Polish dental practices appeal to international travelers by offering comprehensive packages for their trips, covering transportation to and from the airport, lodging, procedures, and aftercare, to start with, making planning a dental tourism trip to Poland both simple and streamlined (Napitu, 2022a).

Throughout their five-year professional education, Polish dentists are required to study and learn dentistry in English, or another relevant foreign language outside of Polish, creating a wealth of dental professionals that can communicate with international dental tourists with ease. Clinics are regulated by the Polish Ministry of Health, while dentists themselves must register with and be regulated by the Polish Chamber of Physicians and Dentists.

Pricing: Dental implants reportedly begin at $700 per tooth in Poland, compared to the $2000 average in the United States, while porcelain veneers begin at $400 per tooth, and full implants at $5200, offering savings to US dental tourists between 50% and 75%.

Spain: The westernmost European country with a bustling dental tourism industry, Spain has become a simple and convenient choice for dental care for people hailing all across Europe, particularly the United Kingdom, and a germane destination for Americans looking to pair dental tourism with a vacation. Held to the standards of the European Union, Spanish dental professionals are known for their quality of care and high level of efficiency, and because there is no public dental system established for adults, the large number of privatized Spanish dentists has created enough competition to keep the average cost of dental procedures down (Napitu, 2022e).

Many dentists in Spain speak English, and the English Speaking Healthcare Association provides a database for potential dental tourists to source an English-speaking dentist that suits their needs (English Speaking Healthcare Association Spain, n.d.). Dental professionals in Spain are required to complete five years of dental education before they can practice, and all are required to join the Spanish Dental Association for regulatory purposes, and the level of modernization and advancement in Spanish dental facilities is up to the strict standards laid out by the European Union.

Beyond dental travel, Spain is the world's second most popular tourism destination in the world, welcoming more than 83.7 million international visitors in 2018 alone (Salvatierra, 2019). With a rich cultural history, vibrant cities to explore like Barcelona and Madrid, beautiful coastlines and world-famous cuisine, Spain is a great option to receive cost-efficient dental care and vacation at the same time.

Pricing: Costs for dental care are nearly half that of the United States in Spain, offering single-tooth implants at $1000 per tooth on average, dental crowns starting at $400, and porcelain veneers at $300 per tooth.

Thailand: Thailand has become an increasingly popular choice for dental tourism in recent years, especially as the Thai government has put a prominent focus on growing the country's tourism sector. The Thai dental sector has become known for its proficiency in cosmetic dentistry, providing competitive pricing at around 60% of the price of care in Western countries, and offers numerous packages for dental tourists to pair their dental tourism with a relaxing vacation in Thailand's renowned tropical backdrop (Thantakit International Dental Center, 2022). Even with the cost of travel factored in, Thailand's cost-efficient pricing models can still allow international patients to save money on the cost of their dental care.

Thailand has promoted the education of new dentists in the country, offering subsidies for the standard five years of dental school so long as dentists serve in the Thai public dental sector for three years before entering private practice, and dentistry is widely regarded across the country as a highly respected professional path. Upon graduation from school, all Thai dentists are automatically granted a license from the country's regulatory dental body, the Thai Dental Council (Thai Dental Council, n.d.). Many Thai facilities are kept up to the highly modernized and technological standards of their Western counterparts, but potential dental tourists to the country would be wise to extensively vet whichever provider they wish to receive care from, as the standards are not always consistently met.

Pricing: In Thailand, single-tooth implants begin at $800 on average, all-on-4 implants at $8000, porcelain veneers at $300 per tooth, and root canals at $140, offering savings above 50% for dental tourists hailing from the United States (Napitu, 2022b).

Turkey: As one of the world's most sought-after destinations for medical tourism, which services more than 500,000 international tourists per year, Turkey's dental tourism sector has been flourishing in its own right. Prices for dental care in Turkey can often be 50–70% lower than the prices for the same procedures in the United States, making it an attractive choice for dental tourists looking for savings. With direct flights entering the country from Europe and Asia, as well as from select airports in the United States, Turkey is an accessible choice for dental tourists across the world. Likewise, Turkey's Mediterranean climate and sightseeing opportunities make Turkey an intriguing option for people looking for a dual-purpose dental tourism visit. Though not a member of the European Union, Turkey's membership in the European Union–Turkey Customs Union allows Europeans to travel into Turkey for dental care with ease, and Turkey offers all of its international tourists a number of medical visas to use to enter the country for the purpose of medical and dental tourism (Delegation of the European Union to Turkey, n.d.; Turkey Ministry of Foreign Affairs, n.d.).

Turkey has maintained a continued focus on developing the standards of its medical and dental sectors to the highest possible standard for some time now and requires all of its dental professionals to attend dental university for five years under regulation of the Turkish Ministry of Health before they are able to practice dentistry professionally. Postgraduation, dentists are required to register with the Turkish Dental Association, which helps continue dental education and ensure regulatory standards are met. Turkey is known for utilizing state-of-the-art technology and equipment throughout its dental facilities, and many hospitals throughout Turkey have received international accreditation and ensure the facilities are up to global standards.

Pricing: Turkey's dental tourism industry allows for patients to receive care for less than half the cost of the United States, with single-tooth implants beginning at $800, dental crowns at $300, porcelain veneers at $200 per tooth, and laser whitening at $200—some $1300 than the treatment costs in the United States on average (Napitu, 2021).

Vietnam: Vietnam is one of Asia's and the world's largest players in the dental tourism industry, bringing in 100,000 international travelers for the purpose of dental tourism and pulling in $150 million in revenue for the country in 2017 alone (Saigoneer, 2018). Ahead of the COVID-19 pandemic, the Vietnamese government projected more than 250,000 international dental tourists to enter the country by the year 2022 and continues to follow through on its goal of becoming an international hub for dental tourism (Hà, 2017; Okhifun, n.d.). Though the pandemic put a brief pause to this upward trajectory, Vietnam is expected to continue solidifying its position as one of the most sought-after dental tourism destinations around the globe in the years to come. Vietnam also hosts cooperation programs with both the United States and Australia, providing additional resources and reassurance for dental tourists originating from these locations (I Tour Vietnam, 2020). However, English proficiency is not common across the Vietnamese dental industry, and finding a provider fluent in English will require additional research or outside assistance.

> There are over 1790 public dental clinics (or health facilities offering dental care) in Vietnam, at least five in each district, while the number rises to over 100 in main cities. The number of private dental clinics is increasing rapidly (there were 70 in 2008, latest data available from Ministry of Health), mainly concentrated in urban areas of Hanoi, Ho Chi Minh City and Da Nang, with only few rural residents having access to any dental health services. Dental equipment is quite entirely supplied by imports as dental equipment manufactured domestically is still limited to furniture and simple equipment. Hanoi and Ho Chi Minh City represent 80% of the entire dental market and demand for emerging dental clinics and replacement of old equipment foresees good market potential for dental devices and supplies. (Borriello, 2021)

Filled with world heritage sites, a friendly population, beautiful natural flora and fauna, and pristine beaches, Vietnam's multitude of attractions has inspired many dental tourists to expand their medical journeys to encompass more traditional tourism elements. To promote this, many dental clinics across Vietnam offer comprehensive packages for dental tourists, providing them with airport transportation, dental consultations and procedures, hotel lodging, and sightseeing tours.

Pricing: Vietnamese dental care is priced significantly lower than in the United States, offering root canals starting at $50, single-tooth implants at $700, all-on-4 implants for $5800, and porcelain crowns for $261 per tooth, or one-third of the price of average US dental care (Medical Tourism Corporation, n.d.).

KEYWORDS

- **medical tourism**
- **travel for treatment**
- **dentistry**
- **dental tourism**
- **oral care**
- **oral health**
- **dental care**

REFERENCES

Aesthetics Dental Care. "5 Things about Dental Tourism in Indonesia." *ADC Dentist*, July 20, 2020. adcdentist.co.id/5-things-you-need-to-know-about-dental-tourism-in-indonesia/

Akbar, F. H.; et al. Dental Tourism: New Strategies for the Health Care in Indonesia. *Int. J. Sci. Technol. Res.* **2020,** *9* (2).

Alvarado, L. How to Receive a Lifetime Guarantee on Dental Work in Costa Rica. *The Costa Rica Star*. n.d. news.co.cr/how-to-receive-a-lifetime-guarantee-on-dental-work-in-costa-rica/74890/

American Cancer Society. *Key Statistics for Oral Cavity and Oropharyngeal Cancers*. Cancer.org, American Cancer Society, 2019, www.cancer.org/cancer/oral-cavity-and-oropharyngeal-cancer/about/key-statistics.html

American Dental Association. *HIV/AIDS and Oral Health*, n.d. www.mouthhealthy.org, www.mouthhealthy.org/en/az-topics/h/hiv-aids-and-dental-health

American Dental Education Association. *Educational Debt*, n.d. www.adea.org, www.adea.org/GoDental/Money_Matters/Educational_Debt.aspx

Asmussen, N. *Why Get Dental Implants in Hungary: Dentists Abroad*, Mar. 6, 2022. Dentaly.org, www.dentaly.org/us/dental-tourism-usa/dental-implants-hungary/ (accessed July 30, 2022).

Association of the Statutory Health Insurance Dentists. *KZBV—Medizinische Infos*, n.d. www.kzbv.de, www.kzbv.de/medizinische-infos.26.de.html#

Bite Magazine. Holiday in Cambodia. *Bite Magazine*, Nov. 19, 2013, www.bitemagazine.com.au/holiday-in-cambodia/

Blog.titanwebagency.com. *Top 20 Dental Industry Trends in 2021 & What They Mean to Practice Growth*. Blog.titanwebagency.com, 2022. blog.titanwebagency.com/dental-industry-trends

Borriello, S. *Growth in Dental Problems, Driving Growth in Vietnamese Dental Industry*; Infodent International, 2021.

Brignall, M. Poland's Medical Tourism Clinics Offer Half-Price Treatment to the World. *The Guardian*, May 31, 2014. www.theguardian.com/business/2014/jun/01/poland-medical-tourism-half-price (accessed July 30, 2022).

CareQuest Institute for Oral Health. Major National Survey Reveals Impending Spike in Oral Health Needs. *GlobeNewswire News Room*, Apr. 8, 2021. www.globenewswire.com/news-release/2021/04/08/2206983/0/en/Major-National-Survey-Reveals-Impending-Spike-in-Oral-Health-Needs.html#:~:text=Based%20on%20the%20survey%20results

Center for Disease Control and Prevention. *Disparities in Oral Health*; Centers for Disease Control and Prevention, 2019a. www.cdc.gov/oralhealth/oral_health_disparities/index.htm

Center for Disease Control and Prevention. *Periodontal Disease*; Centers for Disease Control and Prevention, 2019b. www.cdc.gov/oralhealth/conditions/periodontal-disease.html

Center for Disease Control and Prevention. *Heart Disease Facts*; Centers for Disease Control and Prevention, Dec. 2, 2019c. www.cdc.gov/heartdisease/facts.htm#:~:text=Heart%20disease%20is%20the%20leading

Cleveland Clinic. *Oral Health & Risk for CV Disease*; Cleveland Clinic, n.d. my.clevelandclinic.org/health/articles/11264-oral-health--risk-for-cv-disease

Cooperation Committee for Cambodia. *Cambodian Dental Association*, n.d. www.ccc-Cambodia.org, www.ccc-cambodia.org/en/ngodb/ngo-information/1902 (accessed July 30, 2022).

Cvetanovski, B.; et al. *Growing Faster than the Market: Three Questions the C-Suite Should Ask | McKinsey*, n.d. www.mckinsey.com, www.mckinsey.com/business-functions/growth-marketing-and-sales/our-insights/growing-faster-than-the-market

CzechTourism. *Dentistry*, n.d. www.visitczechrepublic.com, www.visitczechrepublic.com/en-US/14c5471c-fcd0-401f-9a1d-ddf84dbf2067/place/a-dentistry (accessed July 31, 2022).

Da Silva, C. Thousands of Americans Cross Border into Mexico for Affordable Medical Treatment Each Year. *Newsweek*, May 18, 2019. www.newsweek.com/thousands-americans-cross-border-mexico-affordable-medical-treatment-each-1426943

Dalanon, J.; Matsuka, Y. The Evolution of Dental Education in the Philippines. *Health Prof. Educ.* **2019,** *5* (3), 172–176. www.sciencedirect.com/science/article/pii/S2452301118300178?via%3Dihub, doi:10.1016/j.hpe.2018.07.001.

Delegation of the European Union to Turkey. *Customs Union, Supporting the Growth of EU—Turkey Trade*; EU Delegation to Turkey, n.d. www.avrupa.info.tr/en/customs-union-supporting-growth-eu-turkey-trade-52

Deloitte. The Food Service Industry Post-COVID. *Deloitte United States*, n.d. www2.deloitte.com/us/en/pages/consumer-business/articles/food-service-restaurant-business-trends-post-covid.html

Dental Implant Cost Guide. Cost of Getting Dental Implants in Costa Rica. *Dental Implant Cost Guide*, n.d. www.dentalimplantcostguide.com/costa-rica/ (accessed July 31, 2022).

Dental Tourism Expert. *Prague Dentists*; Dental Tourism Expert, n.d. www.dentaltourismexpert.com/prague-dentists/ (accessed July 31, 2022).

Dental Tourism in Colombia. Quality Dentistry Offering up to 95% Savings of the US Prices. *DentaVacation*, n.d. www.dentavacation.com/dental-work-in-colombia/ (accessed July 31, 2022).

Dental Travel Services. *Dental Tourism in Cambodia*, n.d. www.dentaltravelservices.com, www.dentaltravelservices.com/dental-destination-guides/dental-tourism-asia-region/cambodia-dental-tourism

Dentaly.org. *Dental Implants in Mexico and Other Dental Work for Cheap*, n.d. Dentaly.org. www.dentaly.org/us/dental-tourism-usa/dental-implants-mexico/#Mexico_dentist_prices (accessed July 29, 2022).

Dentaround. *7 Things That Make Bulgaria Top Dental Tourism Destination*; Dentaround.com, n.d. dentaround.com/blog/destinations/bulgaria-the-ideal-destination-for-dental-tourism (accessed July 30, 2022).

Dillenberg, J. The Mouth Is the Gateway to the Whole Body. Experts Explain How Oral Health Influences Overall Wellness—Gently by Curaden. *Gently*, Dec. 2, 2019. gently.curaden.com/the-mouth-is-the-gateway-to-the-whole-body/ (accessed July 29, 2022).

Doyle, L. Dental Tourism in Latin America: 5 Top Countries for High Quality Care at Low Cost. *CNBC*, n.d. www.cnbc.com/2017/08/10/dental-tourism-in-latin-america-5-top-countries-for-high-quality-care-at-low-cost.html

Edupark. *Study Medicine and Dentistry in Germany*; Edupark.com.np, n.d., edupark.com.np/undergraduate-studies/study-medicine-in-germany

English Speaking Healthcare Association Spain. *Spanish Directory*; ESHA—English Speaking Healthcare Association: Spain, n.d. eshaspain.org/directory/categories/english-speaking-dentists/ (accessed July 31, 2022).

European Union Cross Border Care Collaboration. *Hungarian Case Study on Dental Care and Patient Flow*, 2013. semmelweis.hu/emk/files/2019/09/Final_case_study_web.pdf

Freed, M.; et al. *Policy Options for Improving Dental Coverage for People on Medicare*; The Henry J. Kaiser Family Foundation, Sept. 18, 2019. www.kff.org/medicare/issue-brief/policy-options-for-improving-dental-coverage-for-people-on-medicare/

Future Market Insights. *Dental Tourism Market Value to Grow by almost US$ 24.3 bn during 2022–2032, Future Market Insights*; GlobeNewswire News Room, May 10, 2022. www.globenewswire.com/news-release/2022/05/10/2439183/0/en/Dental-Tourism-Market-Value-to-Grow-by-Almost-US-24-3-Bn-During-2022-2032-Future-Market-Insights.html (accessed July 29, 2022).

Geib, E. *Dental Services Boom in Mexico*; WEDC, n.d. wedc.org/export/market-intelligence/posts/dental-services-boom-mexico/#:~:text=However%2C%20according%20to%20data%20from (accessed July 31, 2022).

Gheorghe, R.; et al. Dental Tourism from Switzerland to Germany. *Swiss Dent. J.* **2017,** *127* (7–8), 618–633. pubmed.ncbi.nlm.nih.gov/28752506/ (accessed July 31, 2022).

Gibney, M. Healthcare Inequity Creates "Crisis of Trust" among US Marginalized Communities. www.spglobal.com, Oct. 21, 2020. www.spglobal.com/marketintelligence/en/news-insights/latest-news-headlines/healthcare-inequity-creates-crisis-of-trust-among-us-marginalized-communities-60807253#:~:text=Groups%20of%20people%20in%20the (accessed July 29, 2022).

Goel, A. *Dental Tourism in Bulgaria | Safe, Quality, Affordable Dental Works*; www.medicaltourismco.com, n.d. www.medicaltourismco.com/dental-work-in-bulgaria/ (accessed July 30, 2022).

Goodman, P. Costa Rica Dental Prices: How Much Does Dental Care Cost in Costa Rica? *The Tico Times | Costa Rica News | Travel | Real Estate*, Apr. 22, 2020. ticotimes.net/2020/04/22/costa-rica-dental-prices-how-much-does-dental-care-cost-in-costa-rica (accessed July 31, 2022).

Hà, S. *HCM City Keen to Develop Dental Tourism*; Vietnamnews.vn, May 22, 2017. vietnamnews.vn/society/376762/hcm-city-keen-to-develop-dental-tourism.html#EWfmuYeZPpkXByDK.97 (accessed July 31, 2022).

Hamper, A. Medical and Dental Tourism Rising 12%–15% Annually in Central and Eastern Europe. *Prague Post*, June 3, 2018. Hanson, S. 3 Out of 4 Americans Fear Dental Tourism [Survey]. *RetireGuide*, July 22, 2022. www.retireguide.com/guides/dental-tourism/ (accessed July 27, 2022).

Healthinsurance.org. *To What Degree Does Medicaid Provide Adult Dental Coverage?* Healthinsurance.org, May 24, 2022. www.healthinsurance.org/faqs/to-what-degree-does-medicaid-provide-adult-dental-coverage/ (accessed July 29, 2022).

Honova, N. K. Dentistry in Germany. *Expat Guide to Germany | Expatica*, July 26, 2022. www.expatica.com/de/healthcare/healthcare-services/dentists-in-germany-1291023/

I Tour Vietnam. *Dental Tourism in Vietnam on the Rise*; I Tour Vietnam, July 3, 2020, www.itourvn.com/blog/dental-tourism-in-vietnam-on-the-rise

iData Research. *Dental Implants Market & COVID-19 Impact Analysis | Mexico | 2021–2027*; IData Research, Mar. 8, 2021. idataresearch.com/product/dental-implants-market-mexico (accessed July 31, 2022).

Impey, J. *"Dental Tourists" Search for Treatment in Bulgaria*; Deutsche Welle, November 19, 2011. www.dw.com/en/dental-tourists-search-for-treatment-in-bulgaria/a-15534834 (accessed July 31, 2022).

Indian Dental Association. *Careers in Dentistry*; www.ida.org.in, n.d. www.ida.org.in/AboutUs/Details/Careers-In-Dentistry (accessed July 31, 2022).

Indonesia Secretariat of Tourism. *Visa Free for 169 Countries to Travel to Indonesia—Indonesia Travel*, n.d. www.indonesia.travel, www.indonesia.travel/gb/en/trip-ideas/visa-free-for-169-countries-to-travel-to-indonesia#:~:text=All%20travelers%20to%20Indonesia%20must (accessed July 31, 2022).

Jacobs, H. This Tiny Mexican Town Has More than 350 Dentists—and Treats the Thousands of Americans Who Flock across the Border for Dental Care. *Business Insider*, Jan. 2, 2018. www.businessinsider.com/andrew-waits-photos-of-molar-city-2014-6#many-people-travel-to-los-algodones-because-they-dont-have-insurance-and-they-can-get-treatment-and-medication-there-at-a-fraction-of-the-cost-charged-in-america-1 (accessed July 29, 2022).

Jacobs, W. *The Complete Guide to Dental Implants in Colombia*; Apollo Medical Travel, June 14, 2021. apollomedicaltravel.com/2021/06/14/dental-implants-in-colombia/ (accessed July 31, 2022).

Jaramillo, J. A.; et al. Dental Education in Colombia. *J. Oral Sci.* **2010,** *52* (1), 137–143. www.ncbi.nlm.nih.gov/pmc/articles/PMC4090921/#:~:text=To%20obtain%20licensure%2C%20Colombian%20dental. doi:10.2334/josnusd.52.137 (accessed July 31, 2022).

Johns Hopkins Medicine. More Evidence of Link between Severe Gum Disease and Cancer Risk. *Johns Hopkins Medicine Newsroom*, Jan. 18, 2018. www.hopkinsmedicine.org/news/newsroom/news-releases/more-evidence-of-link-between-severe-gum-disease-and-cancer-risk#:~:text=They%20found%20a%2024%20percent (accessed July 29, 2022).

Kaiser Family Foundation. Distribution of Medicare Beneficiaries by Federal Poverty Level. *KFF*, Oct. 23, 2020. www.kff.org/medicare/state-indicator/medicare-beneficiaries-by-fpl/?currentTimeframe=0&sortModel=%7B%22colId%22:%22Location%22 (accessed July 29, 2022).

Karolyi, M.; et al. Mapping of Dental Care in the Czech Republic: Case Study of Graduates Distribution in Practice. In *Proceedings of the 2019 Federated Conference on Computer Science and Information Systems*, Sept. 26, 2019. doi:10.15439/2019f69 (accessed July 31, 2022).

Khazan, O. The Hell of Living without Dental Care. *The Atlantic*, Sept. 25, 2014. www.theatlantic.com/health/archive/2014/09/why-dont-we-treat-teeth-like-the-rest-of-our-bodies/380703/#:~:text=About%20a%20third%20of%20people (accessed July 29, 2022).

Kovacs, E.; Szocska, G. "Vacation for Your Teeth"—Dental Tourists in Hungary from the Perspective of Hungarian Dentists. *Br. Dent. J.* **2013,** *215* (8), 415–418. www.nature.com/articles/sj.bdj.2013.995, 10.1038/sj.bdj.2013.995

Lanmark360. *A Survey of COVID-19 Dental Patients' Perceptions and Attitudes toward Returning to Dental Practices*, n.d.

Live and Invest Overseas. The World's 7 Best Countries for Dental Tourism. *Live and Invest Overseas*, May 9, 2022. www.liveandinvestoverseas.com/health-care/dental-tourism-destinations/ (accessed July 29, 2022).

Lopez, J. "How Costa Rica Became a Top Destination for Dental Tourism." *The Costa Rica Star*, n.d. news.co.cr/how-costa-rica-became-a-top-destination-for-dental-tourism/34058/

Madigan, J. Is Costa Rica a Safe Destination for Dental Tourism? *The Tico Times | Costa Rica News | Travel | Real Estate*, June 26, 2019. ticotimes.net/2019/06/26/is-costa-rica-a-safe-destination-for-dental-tourism (accessed July 31, 2022).

Martin, L. *Diabetes and Teeth—American Dental Association*; Mouthhealthy.org, 2019. www.mouthhealthy.org/en/az-topics/d/diabetes

Medical Tourism Corporation. *Dental Work in Vietnam*; Medical Tourism Corporation, n.d. www.medicaltourismco.com, www.medicaltourismco.com/dental-work-in-vietnam/

Medical Tourism Review. "Dental Tourism in Hungary. Prices and Benefits." *Medical Tourism Review*, Oct. 29, 2020. medicaltourism.review/articles/dental-tourism-hungary-prices-and-benefits (accessed July 31, 2022).

Mendoza, J. *Revenue Medical Tourism Colombia 2024*; Statista, n.d. www.statista.com/statistics/1156551/colombia-revenue-medical-tourism/

Napitu, A. *Dental Implants in Turkey: Costs of Restorative and Cosmetic*; Dentaly.org, Nov. 10, 2021. www.dentaly.org/us/dental-tourism-usa/dental-care-turkey/ (accessed July 31, 2022).

Napitu, A. *Dental Implants in Poland and Other Dental Treatment*; Dentaly.org, July 12, 2022a. www.dentaly.org/us/dental-tourism-usa/dental-implants-poland/ (accessed July 31, 2022).

Napitu, A. *Dental Implants in Thailand and Other Work: Prices and Reviews*; Dentaly.org, Mar. 31, 2022b. www.dentaly.org/us/dental-tourism-usa/thailand-dental-implants/ (accessed July 31, 2022).

Napitu, A. *Dental Implants in the Philippines: Costs & Other Treatment*; Dentaly.org, Jan. 3, 2022c. www.dentaly.org/us/dental-tourism-usa/dental-implants-philippines/ (accessed July 31, 2022).

Napitu, A. *Indian Dental Tourism: Dental Implants Cost in India & More*; Dentaly.org, July 12, 2022d. www.dentaly.org/us/dental-tourism-usa/dental-implants-india/ (accessed July 31, 2022).

Napitu, A. *Spain Dental Tourism: Dental Implants & Other Treatments*; Dentaly.org, Feb. 23, 2022e. www.dentaly.org/us/dental-tourism-usa/dental-implants-spain/ (accessed July 31, 2022).

Nasseh, K.; et al. Consolidation in the Dental Industry: A Closer Look at Dental Payers and Providers. *Int. J. Health Econ. Manage.* **2019,** *20* (2), 145–162. doi:10.1007/s10754-019-09274-x (accessed Dec. 5, 2021).

National Institute of Dental and Craniofacial Research. *Dental Caries (Tooth Decay) in Adults (Age 20 to 64)*; Nih.gov, 2018. www.nidcr.nih.gov/research/data-statistics/dental-caries/adults

Office of Disease Prevention and Health Promotion. *Leading Health Indicators | Healthy People 2020*. Healthypeople.gov, 2014. www.healthypeople.gov/2020/Leading-Health-Indicators

Okhifun, G. *Vietnam: An Emerging Global Leader in Dental Tourism.* www.magazine.medicaltourism.com, n.d. www.magazine.medicaltourism.com/article/vietnam-an-emerging-global-leader-in-dental-tourism (accessed July 31, 2022).

Österle, A.; et al. Travelling for Teeth: Characteristics and Perspectives of Dental Care Tourism in Hungary. *Br. Dent. J.* **2009,** *206* (8), 425–428. www.nature.com/articles/sj.bdj.2009.308/tables/. doi:10.1038/sj.bdj.2009.308.

Pagna Dental. *Price List*; Dental Clinic, n.d. pagnadental.com/price-list/ (accessed July 31, 2022).

Paju, S., Scannapieco, F.A. Oral Biofilms, Periodontitis, and Pulmonary Infections. *Oral Dis.* **2007,** *13* (6), 508–512. www.ncbi.nlm.nih.gov/pmc/articles/PMC2258093/#:~:text=Dental%20plaque%2C%20a%20tooth%2Dborne. doi:10.1111/j.1601-0825.2007.1410a.x.

Pallarito, K. Many Americans Skip the Dentist Due to Cost. *Consumer Health News | HealthDay*, Dec. 6, 2016. consumer.healthday.com/dental-and-oral-information-9/

misc-dental-problem-news-174/many-americans-skip-the-dentist-due-to-cost-717507.html (accessed July 29, 2022).

Peručić, D. Limitations and Development Opportunities of Dental Tourism: The Case of Croatia—ProQuest. www.proquest.com, Feb. 15, 2019. www.proquest.com/openview/ddaede54b9358dd99ed4ffcbab86bc8c/1?pq-origsite=gscholar&cbl=2031970 (accessed July 31, 2022).

Rivera, S. Study Finds Medical and Dental Care South of the Border Are Good Options for Americans. *BorderReport*, July 28, 2021. www.borderreport.com/news/health/study-finds-medical-and-dental-care-south-of-the-border-are-good-options-for-americans/

Robbins, T. "A Reason to Smile: Mexican Town Is a Destination for Dental Tourism. NPR.org, June 9, 2014. www.npr.org/sections/health-shots/2014/06/09/318212444/a-reason-to-smile-mexican-town-is-a-destination-for-dental-tourism

Rosato, D. Think Twice before Buying Private Dental Insurance. *Consumer Reports*, n.d. www.consumerreports.org/insurance/private-dental-insurance-think-twice-before-buying/ (accessed July 29, 2022).

Ruddh, R. "Dental Tourism in India: A Brave New World." *MyVoice*, July 14, 2018. myvoice.opindia.com/2018/07/dental-tourism-in-india-a-brave-new-world/ (accessed July 31, 2022).

Saigoneer. *Vietnam's Emergence as a Global Leader in Dental Tourism | Saigoneer*; Saigoneer.com, Nov. 19, 2018. saigoneer.com/sponsored-listings/14922-vietnam%E2%80%99s-emergence-as-a-global-leader-in-dental-tourism (accessed July 31, 2022).

Salinas, T. J. Your Teeth and Your Heart: What's the Connection? *Mayo Clinic*, n.d. www.mayoclinic.org/healthy-lifestyle/adult-health/expert-answers/heart-disease-prevention/faq-20057986#:~:text=Gum%20disease%20(periodontitis)%20is%20associated

Salvatierra, J. Spain Sets New Foreign Tourist Arrival Record for Sixth Consecutive Year. *El País*, Feb. 4, 2019. english.elpais.com/elpais/2019/02/01/inenglish/1549014848_149720.html

Schrimshaw, E. W.; et al. Insurance-Related Barriers to Accessing Dental Care among African American Adults with Oral Health Symptoms in Harlem, New York City. *Am. J. Publ. Health* **2011,** *101* (8), 1420–1428. doi:10.2105/ajph.2010.300076.

Social Security Administration. "Social Security Act." ssa.gov, Feb. 22, 2018. www.ssa.gov/OP_Home/ssact/ssact-toc.htm

Spector, N. Oral History: Why Teeth Are Treated Different than Our Bodies. *NBC News*, Oct. 24, 2017. www.nbcnews.com/better/health/reason-your-dental-work-isn-t-covered-medical-insurance-ncna813666

Steingold, D. The Sad Tooth: 6 in 10 Adults Are Too Scared to Visit the Dentist. *Study Finds*, Feb. 27, 2018. www.studyfinds.org/adults-too-scared-visit-dentist-teeth/

Thai Dental Council. "ทันตแพทยสภา." *Dentalcouncil.or.th*, n.d. www.dentalcouncil.or.th/en/index.php (accessed July 31, 2022).

Thantakit International Dental Center. *Thailand Dental Tourism: The Ultimate Guide. Thantakit International Dental Center*, May 23, 2022. www.thantakit.com/thailand-dental-tourism/ (accessed July 31, 2022).

Tigg, F. N. R. *Rick Ross Flies to Colombia to Get New Set of Veneers*. sports.yahoo.com, 2020. sports.yahoo.com/rick-ross-flies-colombia-set-171630265.html (accessed July 31, 2022).

TourismReview.org. "Dental Tourism in Costa Rica." www.tourism-review.com, Sept. 2, 2019. www.tourism-review.com/travel-tourism-magazine-dental-tourism-in-costa-rica-article2731

Turkey Ministry of Foreign Affairs. *General Information about Turkish Visas*; Rep. of Turkey Ministry of Foreign Affairs, n.d. www.mfa.gov.tr, www.mfa.gov.tr/general-information-about-turkish-visas.en.mfa

United Kingdom Government. *Money—Cambodia Travel Advice*, n.d. gov.uk, www.gov.uk/foreign-travel-advice/cambodia/money

University of Illinois Chicago. *The Many Costs (Financial and Well-Being) of Poor Oral Health | College of Dentistry | University of Illinois at Chicago*. uic.edu, 2019. dentistry.uic.edu/news-stories/the-many-costs-financial-and-well-being-of-poor-oral-health/

Vijay, N. *India Propels Its Dental Tourism Growth with Advanced Technology & Reasonable Procedural Cost: Dr Meghna Dang*. www.pharmabiz.com, Nov. 23, 2019. www.pharmabiz.com/NewsDetails.aspx?aid=119540&sid=1 (accessed July 31, 2022).

World Health Organization. *Health Workforce: Dentistry Personnel*; www.who.int, 2022. www.who.int/data/gho/data/themes/topics/indicator-groups/indicator-group-details/GHO/dentistry-personnel (accessed July 31, 2022).

www.expat.or.id. *Dental and Orthodontic Care in Bali, Jakarta and Surabaya, Indonesia—Health and Medical Concerns*, n.d. www.expat.or.id, www.expat.or.id/medical/dentalcareinindonesia.html (accessed July 30, 2022).

www.praguepost.com/lifestyle/medical-dental-tourism-rising-in-europe (accessed July 31, 2022).

Yenen, Z.; Ataçağ, T. Oral Care in Pregnancy. *J. Turk. German Gynecol. Assoc.* **2019,** *20* (4), 264–268. www.ncbi.nlm.nih.gov/pmc/articles/PMC6883753/. doi:10.4274/jtgga.galenos.2018.2018.0139.

Yuniar, R. W. Hong Kong Domestic Worker Arrests Casts Spotlight on Indonesia's Self-Taught Dentists. *South China Morning Post*, May 2, 2021. www.scmp.com/week-asia/health-environment/article/3131848/indonesias-self-taught-dental-workers-say-they-fill (accessed July 31, 2022).

CHAPTER 12

RISE IN MEDICAL TRAVEL FOR A SECOND OPINION

ABSTRACT

Patients are now more aware and inclined toward value-based care, and so the medical fraternity is observing an immense rise in patients seeking a second opinion; a huge part of this owes to patients being a victim of misdiagnosis at least once throughout their life, as misdiagnosis has also been a prevalent cause behind approximately 11% of patient deaths. These numbers are devastating and speak loud enough about the unequivocal need to bring the second opinion into the limelight more than ever. This chapter explores the rise of medical travel for a second opinion in the global healthcare sector. It further explores growing technologies like artificial intelligence and their role in healthcare, health insurance for a second opinion, market research in the second opinion market, and upcoming players resolving the accessibility issues to source a second opinion.

In the last few years, patients are becoming more aware and inclined toward value-based care, and so the medical fraternity is observing an immense rise in patients seeking a second opinion. A huge part of it owes to patients being a victim of misdiagnosis themselves at least once throughout their life. According to a study conducted by the National Academy of Medicine, the United States alone sees 6% of patients who are receiving outpatient care being subjected to diagnostic errors. Misdiagnosis has also been a prevalent cause behind approximately 11% of patient deaths. These numbers are devastating and speak loud enough about the unequivocal need to bring the second opinion into the limelight more than ever.

The suspected growth in the second opinion market size in the coming years emphasizes this immense value. The medical second opinion market is

Medical, Dental, and Wellness Tourism: A Postpandemic Perspective. Mary Schreiber Swenson, PhD, and Amit Bansal (Authors).

expected to reach a growth progression worth USD 9751.58 Million by 2027 at a 15.8% CAGR (report by Market Research Future (MRFR)).

It is important to understand that even a second opinion can entail chances of error and therefore a patient can only be benefitted from a second opinion if it is of equal or better quality than the first diagnosis (Hillen et al., 2017). Currently, 10–15% of diagnostic errors are found to be in general medicine as a lack of consulting with the right specialist. This can be easily reduced by receiving more specialized treatment. That is not all, a second opinion provides reassurance to the patients helping them take control of their health at the right times and hence feel psychologically satisfied. (Axon et al., 2008) Dissatisfaction in the patient–physician communication was also a major reason behind many patients opting for a second opinion.

The shift is also visible among the physicians, as they actively prefer seeking a colleague's opinion or may refer patients to a specialist to eliminate any doubt in the diagnosis before beginning treatment (Tam et al., 2005). This trend is however not prevalent in every country around the world. For instance, seeking a second opinion in Korea is criticized and doctors can be seriously offended by the practice. On the other hand, in countries like India and Turkey, it is a fairly common practice.

Some of the barriers that stop patients from seeking a second opinion service may include a sense of shock, information overload, fear, the pressure of losing out on valuable time, and worry of jeopardizing their relationship with their current physician (Peier-Ruser and von Greyerz, 2018). It also falls on the healthcare institutions and doctors to ensure a patient feels unhurried and calm enough to seek a second opinion, if they wish to, especially in conditions like Cancer or before undergoing serious surgeries. Doctors may also schedule follow-ups with the patient to help them deal with the process and help in processing additional information and insights related to their condition.

Health insurance policies are also actively including second opinion programs as a way to cost containment to eliminate and reduce medical bills by discarding any wrongful or ineffective treatment suggestions (Rosenberg et al., 1989; Odeyemi and Nixon, 2013). Hospitals also tend to suggest many screening and routine pathology tests from time to time to review and confirm a diagnosis. Patients also actively seek opinions elsewhere when they are unhappy with the first opinion or results. There has been a gradual change in the diagnosis and suggested line of treatment when shifted from doctor to doctor. Trends have been shifting country-wise as some countries like Japan are equipped with highly modern medical equipment that eliminates the chance of diagnostic error to a large extent, whereas countries like Africa are still dealing with a lack of medical technologies and diagnostic approaches. This encourages the "travel for treatment for obtaining a second opinion." Pathology and radiology play a crucial role in determining the value of a second opinion by producing reports that assist the doctors in forming an opinion (Moumjid et al., 2007; Davidov et al., 2010). In many cases, misdiagnosis occurs as physicians give opinions based on the diagnostic reports instead of actual scans and other processed imaging reports. Just by documenting the radiological and pathological tests, there can be a hike in quality of care and a reduction in diagnostic error.

On the basis of the above description, there are three main types of distinguishing in the second opinion. The first one highlights a patient's need to confirm their diagnosis, treatment, and medical journey as suggested by the first physician. The second one is physician initiated in search for advice from the second expert to weigh on the diagnosis. And the third one is imposed by third-party insurers on both patients as well as doctors as a method of cost containment (Tam et al., 2005).

12.1 INSURANCE COVERAGE FOR A SECOND OPINION

In the 1970s, insurance companies in the United States started realizing the value of a second opinion for patients, especially suffering from critical conditions, and after careful analysis of the changing scenarios, they included second opinion programs in their plans as a pre-authorization tool before elective surgery. The reason remained to be cost containment. With the advent of a second opinion, many critical surgeries are prevented which is a huge cost-saving factor for insurance firms. Still, major differences in the access to healthcare and health policies can be observed from one state to the other in the United States. The right to obtain a second opinion was passed in the states of Florida, Indiana, Louisiana, Missouri, New Hampshire, and New York in the 1990s (Wagner and Wagner, 1999). Even though the plans are sometimes still designed in a way that restricts the patients to obtain a second opinion independently as they necessitate a referral from their primary care physician and allow patients to only see a doctor within their network for a second opinion (Medicare). On the other hand, if you travel to Canada for a second opinion, there is no mandatory requirement for surgery (Ridic et al., 2012). Therefore, travel for treatment also plays a freeing role in the life of patients who are feeling stuck within a restricted healthcare system.

In countries like Europe and the UK, obtaining a second opinion still is not a legal right (NHS Choices, 2012). A primary care physician can simply refuse to recommend their patients for a second opinion if they wish to do so. The problem is evidently in the tightly bounded system that does not allow patients to consider all their options or refer to doctors till the point they reach satisfaction. As a result, the chances of being a victim of misdiagnosis increase leading to severe consequences on the health.

12.2 INFORMATION UPHILL AND SECOND OPINION

Another reason why it is important to consider a second opinion is the increasing instances of "Web Babel Syndrome," in which people visit and refer to the information available on health websites as their first diagnosis. A survey conducted in Europe and the United States targeting 4530 people highlighted that 32% of Europeans and 43% of Americans prefer using online portals and health websites like BBC, Google, Yahoo, etc. to obtain health information (Eaton, 2002). As internet sites, the information available on them is usually for information purposes only and lacks much clinically adequate information, complications, and aftercare information (Nassab et al., 2010). The continuous and compulsive behavior of taking the first opinions from internet information can be partly due to the lack of accessibility to healthcare everywhere, or the high treatment costs. There is a frequent gap in communication between physicians and patients due to this syndrome, especially when the patient is suffering from multiple pathologies that exist in synchrony. Misleading information, prescriptions, and the risk of confusion in treatment can occur as a result (Di Cerbo et al., 2014; Palmieri et al., 2011). Technology can solve this problem by offering virtual consultations with doctors. Online doctor consultations have seen a great rinse since and after the COVID-19 pandemic. There are many renowned players utilizing the advent of technology in favor of the patients by creating software and tools that connect them with doctors and provide them with the comfort and benefits they seek from their homes. ThinkTwice by MediGence is one such innovative and technologically rich platform where scanned reports, DICOM scans, patient history, and all other details can be uploaded hassle-free. ThinkTwice has a Global panel of certified doctors ensuring you only get the best diagnosis and details about the line of treatment. By adding a virtual second opinion service to their health plan, a lot of patients will feel free and out of economic dilemmas to visit a doctor and take their opinion after taking the first opinion from the web.

Additionally, the healthcare system in many countries offers initial patient consultations with a general physician and does not give the option to visit a specialist directly, or even visit another doctor for a second opinion straight away. A person has to get a referral from the primary physician to take a second opinion. Like in the Dutch healthcare system, patients have to cross the general practitioner who acts as a general gatekeeper, to specialist care. Only after the patients have received primary care in a regular hospital they can legally demand a second opinion service. This second opinion is covered by insurance in the Dutch healthcare system (Wet op de geneeskundige behandelingsovereenkomst, n.d.). When a patient demands a second opinion service they are recommended to some other hospital by their current general physician. In this other hospital, a physician is assigned to the patient at similar levels and diagnostic resources that were available in the previous hospital. To brief the same, the level of care during the first and second opinions is carried out at the same level.

However, the whole purpose of second opinion service should be to provide enhanced and better care. Therefore, to take it one notch ahead, and solve this problem, platforms like ThinkTwice stepped up in the healthcare sector using advanced technology. Patients can seek the opinions of experts with years of experience treating patients with similar conditions from all around the world. A panel comes together to discuss the case reports, diagnose the condition and deliver the best line of treatment to the patient within stipulated days. And everything happens smoothly and virtually.

12.3 ARTIFICIAL INTELLIGENCE IN HEALTHCARE

In the past few years, artificial intelligence has taken over every sector, every landscape, and healthcare is no different. Recently, the healthcare fraternity has adopted many AI-based techniques and bots, both in the initial line of diagnosis and to deliver a second opinion. This has also led to the question of whether AI can replace human physicians in the near future. We believe the answer is *No*. No matter the extent of growth in the technology, a human physician will always provide an unparalleled value that can never be fully duplicated by a machine in the foreseeable future. Artificial intelligent bots are trained to assist patients by understanding their behavior, studying thousands of similar as well as unique patient cases, and medical histories, and then adapting to the cases in order to deliver the right diagnostic and treatment plans. Healthcare data is increasingly becoming overwhelming and with the help of big data analytic methods, now it is possible to use applications of technology in certain areas like radiology more effectively than others. AI guided by the right clinical question can look through all the relevant clinical information from a fed massive amount of health data to assist the physician in making the right clinical decision (Murdoch and Detsky, 2013; Kolker et al., 2016; Dilsizian and Siegel, 2014).

Although one important consideration that researchers and medical experts have to keep in mind is that every AI has to be trained through a lot of generated data-like screening, diagnosis, treatment assignment, and so on, before they can be used as a tool for medical purposes. This clinical data can be often found in the form of medical notes, electronic recordings taken by a medical device, the demographics, physical examinations, medical history, laboratory tests, and imaging (Administration UFaD, 2013). And even then, there are ethical issues and a margin of error associated with results produced by AI. However, many researchers suggest doctors combine their efforts with AI-based diagnostics, especially in the case of large data and imaging data containing information. As AI technology has the capacity to perform tasks that mimic human intelligence, the impact on the masses will be huge once it is in practice worldwide. The whole landscape of medicine as currently known to us will shift. Right now trends are showing upward growth as the enthusiasm for AI in healthcare is at its peak with success in radiological image analysis, diagnostic tools, and ChatBots (Matheny et al., 2020; Topol, 2019; He et al., 2019). One major concern of the experts is to safely and efficiently implement AI technologies in healthcare. When in healthcare, the matters can quickly escalate to life and death issues. On the same note, responsible practice and setting key limitations are extremely crucial in healthcare when it comes to technology implementation of any kind.

Moreover, AI can have plenty of benefits in the healthcare field but when you take a magnified glance at the values, priorities, and needs of a patient, it may lack in certain aspects to provide the human touch and intelligence. Patients are the end user of every upcoming technology and they are not open to experimenting with their health in terms of new innovations that pose risks. Implementing the right care and following the ethical guidelines while doing the same is a concern of researchers too. Even if the technology is not interfering with the reports produced for a patient, in case any risk arises patients will be the first party to get most affected by the unethical or incorrect data produced or shortlisted by an AI (Yarborough et al., 2013). The extent is so high that all the repercussions must be accepted by a patient alone associated with the novel application of AI in healthcare. Patient interaction, acceptance, and awareness of medical innovation and technologies are therefore one of the most crucial elements if the healthcare community wants to incorporate AI in an ethical manner (Borry et al., 2004).

Patient concern about AI in healthcare is a significant barrier to the use of these tools. Nonmedical applications in AI have been studied and the results show variable views of the public toward the use of these tools (Fast and Horwitz, 2017). The margin of error, even though it is less to date, is still not trusted by everyone. It is crucial to spread awareness among the masses regarding the key role played by technology in shaping healthcare for the better. Without patient engagement, new progress will be null and void and public trust in healthcare will shatter. This fall in the patient trust can be a reason behind the beginning of another AI winter (Muthukrishnan et al., 2020).

12.4 REASSURANCE AND SAFETY

Real-life stories of people who have benefitted from a second opinion are indeed an eye-opener. We have come across and met with many patients who avoided undergoing the wrong treatment by availing of the second opinion services. For privacy reasons, we cannot disclose their identity. Here's their experience and how a lot of emotional and physical turmoil was ruled out with help of confirmation on the primary diagnosis.

The first patient (Patient 1) was diagnosed with a worn and damaged knee. The knee pain was severe and being an athletic person, there were a lot of motion restrictions that were limiting the day-to-day lifestyle of the patient. The pain was unbearable. The primary physician recommended a knee replacement treatment. Like any surgery, it was a huge decision and a critical one to make. The patient was devastated by this information. It was at that time that someone suggested he opt for a second opinion. He reached out to a renowned doctor and went through all the tests and procedures again. The second opinion revealed very different end results. The patient was asked to follow a rehabilitation routine and the need for a knee replacement was completely ruled out. The unmatchable value of a second opinion is reflected in the way it helps the patient save a lot of precious time, follow the right line of treatment, and also save a lot of money.

In the case of patient 2, there's another eye-opening story that highlights the dire need of taking a second opinion before undergoing any critical procedure. A patient was suffering from some liver-related issues and was about to be discharged from the hospital when the primary care physician declared that he was suffering from final-stage liver cirrhosis. The news came as a surprise to the patient as well as his whole family. They were looking for the most suitable and effective line of treatment when someone suggested the family opt for a second opinion. The patient visited another liver specialist and to everyone's surprise, he was not suffering from liver cirrhosis but a case of liver damage that could be easily treated by some medicines. The patient was then put on the right line of treatment and was discharged completely healthy after a period of three months.

These case studies are proof that a second opinion can save a lot of time and save a patient from undergoing the wrong treatment. If you think deeply about the issue, imagine the loss a patient will face if they are not guided toward the right treatment at the right time. Healthcare is a field where every matter is fragile and may point to life and death. Being careful is always the best policy especially when it comes to critical procedures.

12.5 EXPECTED FUTURE TRENDS

In the period 2021–2028, the expected forecast for growth of the medical second opinion market as per a study conducted by Data Bridge Market Research is at a value of 10,002.27 billion and the growth is a CAGR of 10.67%. The emphasis is shifting in the healthcare market toward a superior treatment that is highly driven by outcomes achieved by obtaining a medical second opinion. The future trends will majorly see a rise in awareness among the people for second opinion services, informed choices leading to cost optimization as a vital factor behind market growth escalation, and penetration of modern technologies and superior treatments around the globe. Most of the factors mentioned above can be owed to the dynamically changing lifestyle, preferences, and needs of the global population. Patients consider "Travel for Treatment" as a necessity, not a liability, even if it is just to obtain a second opinion. Other than that, there is a rapid increase in the number of hospitals and the facilities they are delivering to foreign patients.

Moreover, in the forecasted period 2021–2028, an expected increase in technologies like telemedicine will make availing a second medical opinion from anywhere around the world easier for patients and create new opportunities in the healthcare market. Lack of growth is still one factor that can pull back this growth by limiting it. By spreading knowledge in the backward areas of the world, this obstruction can be eliminated.

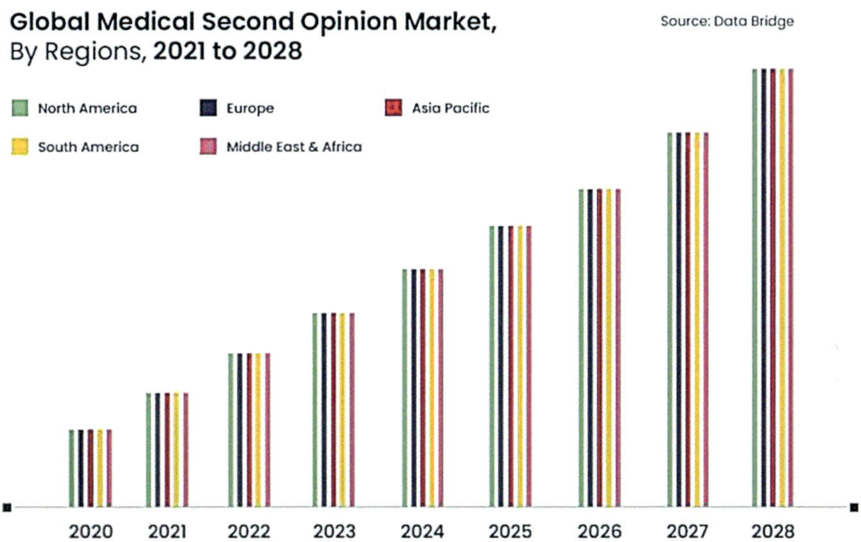

The current market for a medical second opinion is segmented on the basis of service providers, products, and disorders. The sector of the medical second opinion market is divided into the following segments:

- On the basis of disorders: cancer, diabetes, neurological illnesses, blood disorders, cardiac disorders, nephrological disorders, organ transplants, major trauma, and others.
- On the basis of service providers: online services, hospitals, health insurance companies, and others.
- On the basis of product: online second opinions and physical second opinions.

Neurology care, cardiology care, and oncology care are the three most prevalent medical streams where patients actively seek a second opinion. However, in the future, we strongly expect to see growth in other care fields. The current trends show North America to be leading in the second opinion market as the country inhabits a highly aware population with an advanced healthcare system and rising initiatives by the government that favors a second opinion. The presence and availability of global brands in a country also impact the expected growth of the second opinion market.

KEYWORDS

- **medical travel**
- **healthcare tourism**
- **misdiagnosis**
- **second opinion**
- **artificial intelligence**
- **technology**

REFERENCES

Administration UFaD. *Guidance for Industry: Electronic Source Data in Clinical Investigations*, 2013. https://www.fda.gov/downloads/drugs/guidances/ucm328691.pdf (accessed Jun. 1, 2017).

Axon, A.; Hassan, M.; Niv, Y.; et al. Ethical and Legal Implications in Seeking and Providing a Second Medical Opinion. *Dig. Dis.* **2008,** *26*, 11–17. doi:10.1159/000109379.

Borry, P.; Schotsmans, P.; Dierickx, K. What Is the Role of Empirical Research in Bioethical Reflection and Decision-Making? An Ethical Analysis. *Med. Health Care Philos.* **2004,** *7*, 41–53.

Davidov, T.; Trooskin, S. Z.; Shanker, B.-A.; et al. Routine Second-Opinion Cytopathology Review of Thyroid Fine Needle Aspiration Biopsies Reduces Diagnostic Thyroidectomy. *Surgery* **2010,** *148*, 1294–1301. doi:10.1016/j.surg.2010.09.029.

Di Cerbo, A.; Pezzuto, F.; Laurino, C.; Palmieri, B. Web Babel Syndrome and False Expectations During Own Multimedia Oncological Search. *Clin. Ter.* **2014,** 165, e225–e229.

Dilsizian, S. E.; Siegel, E. L. Artificial Intelligence in Medicine and Cardiac Imaging: Harnessing Big Data and Advanced Computing to Provide Personalized Medical Diagnosis and Treatment. *Curr. Cardiol. Rep.* **2014,** *16*, 441. doi:10.1007/s11886-013-0441-8.

Eaton, L. News Roundup: A Third of Europeans and Almost Half of Americans Use Internet for Health Information. *Brit. Med. J.* **2002,** *325*, 989.

Fast, E.; Horvitz, E. In *Proceedings of the Thirty-First AAAI Conference on Artificial Intelligence*; AAAI Press, 2017; pp 963–969.

He, J. X.; et al. The Practical Implementation of Artificial Intelligence Technologies in Medicine. *Nat. Med.* **2019,** *25*, 30–36.

Hillen, M. A.; Medendorp, N. M.; Daams, J. G.; et al. Patient-Driven Second Opinions in Oncology: A Systematic Review. *Oncologist* **2017,** *22*, 1197–1211. doi:10.1634/theoncologist.2016-0429.

Kolker, E.; Özdemir, V.; Kolker, E. How Healthcare Can Refocus on Its Super-Customers (Patients, n=1) and Customers (Doctors and Nurses) by Leveraging Lessons from Amazon, Uber, and Watson. *Omics* **2016,** *20*, 329–333. doi:10.1089/omi.2016.0077.

Matheny, M.; Israni, S. T.; Ahmed, M.; Whicher, D. Artificial Intelligence in Health Care: The Hope, the Hype, the Promise, the Peril. *Natl. Acad. Med.* **2020,** 94–97.

Medicare.gov M go. *Second Opinions before Surgery* [Internet]. 2017 [cited Jun 13, 2017]. Available from: https://www.medicare.gov/what-medicare-covers/part-b/second-opinions-before-surgery.html

Moumjid, N.; Gafni, A.; Bremond, A.; et al. Seeking a Second Opinion: Do Patients Need a Second Opinion When Practice Guidelines Exist? *Health Policy* **2007,** *80*, 43–50. doi:10.1016/j.healthpol.2006.02.009.

Murdoch, T. B.; Detsky, A. S. The Inevitable Application of Big Data to Health Care. *JAMA* **2013,** *309*, 1351–1352. doi:10.1001/jama.2013.393.

Muthukrishnan, N.; et al. Brief History of Artificial Intelligence. *Neuroimag. Clin. N. Am.* **2020,** *30*, 393.

Nassab, R.; Hamnett, N.; Nelson, K.; Kaur, S.; Greensill, B.; et al. Cosmetic Tourism: Public Opinion and Analysis of Information and Content Available on the Internet. *Aesthet. Surg. J.* **2010,** *30*, 465–469.

NHS Choices. *How Do I Get a Second Opinion?—Health Questions—NHS Choices* [Internet]. 2012 [cited Feb 18, 2013]. Available from: http://www.nhs.uk/chq/Pages/910.aspx?CategoryID=68&SubCategoryID=156

Odeyemi, I. A.; Nixon, J. The Role and Uptake of Private Health Insurance in Different Health Care Systems: Are There Lessons for Developing Countries? *Clinicoecon. Outcomes Res.* **2013,** 5, 109–118. doi:10.2147/CEOR.S40386.

Palmieri, B.; Iannitti, T.; Capone, S.; Fistetto, G.; Arisi, E. Second Opinion Clinic: Is the Web Babel Syndrome Treatable? *Clin. Ter.* **2011,** *162*, 575–583.

Peier-Ruser, K. S.; von Greyerz, S. Why Do Cancer Patients Have Difficulties Evaluating the Need for a Second Opinion and What Is Needed to Lower the Barrier? A Qualitative Study. *Oncol. Res. Treat.* **2018,** *41*, 769–773.

Ridic, G.; Gleason, S.; Ridic, O. Comparisons of Health Care Systems in the United States, Germany and Canada. *Mater. Socio-Med.* **2012,** *24*, 112. doi:10.5455/msm.2012.24.112-120.

Rosenberg, S. N.; Gorman, S. A.; Snitzer, S.; et al. Patients' Reactions and Physician-Patient Communication in a Mandatory Surgical Second-Opinion Program. *Med. Care* **1989,** *27*, 466–477. doi:10.1097/00005650-198905000-00003.

Tam, K. F.; Cheng, D. K. L, Ng, T. Y.; et al. The Behaviors of Seeking a Second Opinion from Other Health-Care Professionals and the Utilization of Complementary and Alternative Medicine in Gynecologic Cancer Patients. *Support Care Cancer* **2005,** 13, 679–684. doi:10.1007/s00520-005-0841-4.

Topol, E. J. High-Performance Medicine: The Convergence of Human and Artificial Intelligence. *Nat. Med.* **2019,** *25*, 44–56.

Wagner, T.; Wagner, L. Who Gets Second Opinions? *Health Aff. Proj. Hope* **1999,** *18*, 137–145. doi:10.1377/hlthaff.18.5.137.

Wet op de geneeskundige behandelingsovereenkomst. Burgerlijk Wetboek, Boek 7, n.d.

Yarborough, M.; et al. Relationships Hold the Key to Trustworthy and Productive Translational Science: Recommendations for Expanding Community Engagement in Biomedical Research. *Clin. Transl. Sci.* **2013,** *6*, 310–313.

CHAPTER 13

PRIVACY AND HEALTH DATA SECURITY

ABSTRACT

It is crucial to protect every shred of health data, considering the information included in healthcare documentation is deeply personal and is owned by a healthcare facility but the information it entails is owned by the patient making it a two-key privacy-enabled piece of information. The collection and storage of this information is a major concern as it contains personally identifiable data; in the case of a security breach, there can be severe harm to the individual. In this chapter, the goal of security in healthcare is outlined in a threefold structure: (1) Availability of stored data to only authorized personnel, (2) Restricted access to data, ensuring it is released only for authorized purposes, and (3) Maintaining the accuracy of the information. The chapter also compares HIPAA and GDPR compliances, privacy infrastructure and systems of various countries, and has a checklist for patients to make sure their health data is secure.

A clinical record is a form of a communication tool between a patient and a physician. Electronic or paper-based, these clinical notes later support the process of delivering the right care to the patient by helping in efficient decision-making and supporting evaluation of efficacy and quality of care. The clinical reports also support the coordination of services, legal protection, regulation, and research. A patient's medical history will remain important throughout their life and is critical for all decision-making about current and future treatment possibilities. The medical records are business documents for a healthcare service provider where all of the patient's activities, administrative work, tests, medications, and everything else is documented and

Medical, Dental, and Wellness Tourism: A Postpandemic Perspective. Mary Schreiber Swenson, PhD, and Amit Bansal (Authors).

stored either in a physical or electronic manner. The storage of the clinical records of patients in the electronic format is also referred to as EHR (electronic or digital health record).

Historically, many healthcare institutions are used to keeping healthcare records as a paper repository. This information was readily available and used as a way to keep track of patients' research, administrative, financial as well as clinical details. The data were kept secured even then and limited to one user at a time. The information was updated and maintained manually which lead to many disadvantages like delay in maintaining up-to-date records. Information sometimes lagged behind for months. Most of the healthcare facilities used to keep these records stored in the facility basement as the paperweight was a big burden and they used to take up a lot of space that could be better utilized for other purposes. It was the doctor/physician's responsibility to adhere to data filing and maintaining the care documents. Even the release of any file was preauthorized by the physicians. However, as the patients were less aware then, there was rarely any demand for these medical documents. Another limitation of this paper-based medical record system was a severe lack of security. All information was physically available, just away protected by a few doors and locks. Any one of the authorized users (hospital staff) could easily access the records by using identification cards and sign-out procedures. On the other hand, in most facilities, there were no security measures to highlight unauthorized access to the patient's information. Anyone could view patient reports and no one would know. Taking the case of receipts and paper prescriptions, a patient might be asked to take it with them everywhere they go for a further doctor consultation, diagnostic tests, or a second opinion. This means that a physical copy of patient information will be transferred to many hands. A lot of waste material from clinics and diagnostic centers consists of discarded patient reports and copies of documents. This is a serious problem highlighting the fact that patient data are not really secure in paper document form. Instead, it may be lingering around in hands that are not really keeping it secure.

Jumping to today, the primary concern regarding documentation is still the same to assist in patient care. However, now the clinical documentation is immediately uploaded on electronic systems to make the data compilation process smoother and keep the data readily available in one location till the time the patient is discharged. Every healthcare facility has to meet a stipulated time in which they have to update the patient data and clinical records in the electronic system to meet the regulatory and accreditation criteria. Electronic health records are exclusive yet interactive such that they are available to the reviewers (in the case of healthcare—the doctors) as well as the users. They are easy to access yet secure and password protected. In certain cases, government hospitals have third-party contracts with healthcare agencies that will review patient documentation of care. One benefit that electronic health records provide patients with is that they can be viewed at any time and all the personal health data can be compiled and viewed in a single spot. The information technology tools help in keeping the information about their diagnoses and treatment protected and secured.

13.1 HEALTHCARE DATA PRESERVATION AND DISCLOSURE LIMITATION

Every piece of information that is produced in a healthcare facility while a patient is receiving care, be it a physical record, receipt, or bill is owned by that organization as it is a piece of information that adds to their business record and data. However, the information included in the healthcare documentation is deeply personal and belongs to the patient (Odom-Wesley et al., 2009). The document is owned by the facility but the information it entails is owned by the patient making it a two-key privacy-enabled piece of information. The Office of the National Coordinator for Health Information Technology refers to the health record as "not just a collection of data that you are guarding—it's a life" (Office of the National Coordinator for Health Information Technology, 2012). The three keys ethical priorities revolving around electronic health records are as follows:

1. Privacy and confidentiality
2. Security and data integrity
3. Availability

Richard Rognehaugh defines privacy as "the right of individuals to keep information about themselves from being disclosed to others; the claim of individuals to be let alone, from surveillance or interference from other individuals, organizations or the government" (Rognehaugh, 1999). All of the information that is shared as a result of clinical exchange and the relationship between doctor–patient is confidential data and must be treated that way (Rinehart-Thompson et al., 1999). Sometimes the health information will be stored in the form of videos and other online multimedia files. In the case of recorded telehealth sessions or doctor consultations that are recorded for lifestyle improvement sessions like nutrition or physiotherapy, the question arises "How safe are they, and can I rely on these channels?" This concern for patients has not gone unseen by the providers and healthcare policymakers. The personal information related to patients cannot be released in America without their permission. However, generic hospital information and statistics like the total number of patients with breast cancer that were treated in a hospital from 2019 to 2021 can be disclosed as they do not fall in the data privacy category (Rinehart-Thompson et al., 2006). Though laws must be followed along with patient information for sharing their health data, the rules are not the same for doctors. A physician can gain access to a patient's medical reports for the purpose of treatment, billing, or other administrative tasks without asking for their permission. On the other hand, patients also have the legal rights, both federal and state, to edit any information in their health records or ask for a copy of their medical reports.

So when we talk about disclosure limitation, we are essentially talking about limiting the authority and controlling the access to who can view the medical information. For instance, when we talk of general practitioners, the administrator identifies the user, finds out the level of patient information, and then assigns a password and username to their health data. In many healthcare practices, there is a two-step authentication system in place such as a biometric scan or finger/retina/face scan to ensure the data are highly protected. Additionally, the access is limited to users based on their roles. For instance, a receptionist has no authority to have a look over your personal health data in most cases but your nurse practitioner may need to have a look at the data to deliver the right care for your condition. Determining user privileges is an important aspect of maintaining medical record security. The patient always has access to all the information as it entails their personal data. The healthcare staff is accountable for the use and misuse of patient information and thus must fulfill their role after careful consideration of the fact that their view and any change in the personal health information of patients makes them answerable (American Health Information Management Association, 2012).

13.2 THE HEALTH INSURANCE PORTABILITY AND ACCOUNTABILITY ACT SECURITY RULE AND IMPORTANCE OF PRESERVING HEALTH DATA

It is crucial to protect every shred of data in healthcare. The collection and storage of the pile of information is a major concern as it contains personally identifiable and hence sensitive, at times embarrassing, information. In case of a security breach, there can be severe harm to the individual whose healthcare files have been inappropriately accessed. A patient may suffer intrinsic harm simply because of the disclosure of personal health information to strangers (Saver, 2006). There is also a concern of economic harm in case of health information disclosure. When the sensitive health data becomes public, people have the risk of losing their jobs, housing and face other repercussions due to it. This might also cause social and psychological harm to the individuals. For example, when the information that a person has HIV becomes public, it leads to social isolation and other psychological effects on the patient (Gostin, 2008). Additionally, a security breach of the medical documents also means leakage of all the information that can make identity theft possible.

As briefly mentioned before in this chapter, the goal of security in healthcare can be outlined in a threefold structure:

1. Availability of stored data to only authorized personnel
2. Restricted access to data, ensuring it is released only for authorized purposes
3. The information must be accurate

These objectives have traditionally been met through safeguards designed to keep data processing secure against unauthorized access, change, deletion, or transmission. Healthcare providers must take care to manage patient data appropriately in order to foster a culture of trust and openness with patients while adhering to severe regulatory and privacy requirements. With regularly updated laws and regulations, data privacy in healthcare is a subject that is continually changing. Patients are given the data privacy they expect and deserve in this way.

Over the past several years, new and amended data privacy regulations have been passed in the United States, raising the bar for data privacy across all practices. The Health Insurance Portability and Accountability Act (HIPAA) is the most prominent and well-known law that governs the healthcare sector.

The 1996 law known as HIPAA established the standards for data privacy and security for PHI (protected health information), or health information that needs to be safeguarded. Hospitals, partners, and associated subcontractors are among the businesses that must apply all security measures in order to be HIPAA compliant when handling PHI. Healthcare firms must stay up to date to remain compliant with HIPAA, which is continuously revised to prevent possible threats to data privacy.

Identifiable health information is the data along with demographic information that talks about the past, present, or future health condition, physical or mental, of an individual. It also entails details about the payment provision of an individual in past, present, or future that can make the identity of an individual identifiable or give a strong hint and reason to believe that the information can be used to identify and locate the individual.

The HIPAA privacy rule is applicable to all the individual organizations that are transmitting health information electronically in their everyday healthcare practice. HIPAA also covers any form of written, paper, and spoken healthcare information no matter the size of the information. Other covered entities under HIPAA privacy rule are health plans, healthcare clearinghouses, and healthcare providers. Under the health plan category falls Medicare, Medicaid, or Veterans Affairs, they are either private insurers, government payors, or managed care organizations. On the other hand, healthcare clearinghouses are the institutions that are responsible for billing services. Healthcare providers entail doctors, hospitals, clinics, and other facilities that provide treatment.

An entity that satisfies one of the criteria for a covered entity and undertakes tasks unrelated to healthcare might change its status to a hybrid entity by identifying in writing its "health care components." The Privacy Rule is then only applicable to certain healthcare components. For instance, if an academic medical center and hospital are part of a university, the entire institution will be considered a covered entity unless the university chooses to be a hybrid organization by designating solely the hospital as the healthcare component. As a result, only the hospital is subject to the Privacy Rule. This classification is dependent on the nature of work that is being performed in the organization. If they are dealing with healthcare data via electronic modes then it will be a component of HIPAA (HHS). When owned or maintained by a company other than a covered entity that falls under the HIPAA rule, personally identifiable health information is not protected by the Privacy Rule (HHS, 2007).

13.3 PERMISSIONS FOR USE AND DISCLOSURE UNDER HIPAA

As a general rule, PHI cannot be disclosed without the authorization of an individual whose information is contained in the documents. A covered entity can however use the information without permission for operational and billing-related purposes.

The Privacy Rule allows disclosure of personal health information without the need to ask for a patient's authorization in the following cases:

- Audits, inspections, administrative proceedings, and other essential healthcare oversight activities
- Public health requirements as stated by state and federal law
- To business associates
- Officials in the law enforcement
- In case of a court order for a judiciary hearing
- For purpose of research

Additionally, the Privacy Rule also gives individuals rights regarding their PHI (Pritts, 2008). Individuals have the following rights under the Privacy Rule:

- Get notice of privacy policies from a healthcare provider or a health plan that will advise patients about the planned uses and disclosures of their health information that may be done without the patients' agreement or authorization.
- Request for their personal health information to be viewed and copied.
- Submit a change request for any information that is incorrect or incomplete.
- Obtain a list of the disclosures of PHI made by the covered entity during the previous six years.

13.4 UNDERSTANDING GDPR FRAMEWORK

The General Data Protection Regulation (GDPR) of the European Union provides a comprehensive framework for all types of personal data and adds new concepts to data like health or research. The GDPR treats health data as a special category of data that is sensitive by nature, along with a number of other types, and the following statement is made in this regard: "Processing of personal data revealing racial or ethnic origin, political opinions, religious or philosophical beliefs, or trade union membership, and the processing of genetic data, biometric data for the purpose of uniquely identifying a natural person, data concerning health or data concerning a natural person's sex life or sexual orientation shall be prohibited" (Article 9, section 1; General Data Protection Regulation, The European Union). This quotation illustrates two important points. First off, it is obvious that any of these several sensitive sorts of data could be revealed by social media data. Second, the phrase "data affecting health" is used rather than the actual term "health data." This makes it possible for implied and indirect health data to be covered under the GDPR.

The granting of the person's explicit consent is one of the exceptions that allow for the processing of health data, which is otherwise banned. When it has been explicitly given and the reason for processing the data has been explicitly stated, the collecting of consent from the data subject continues to be one of the most frequent exceptions that companies processing health data will be able to depend on.

Comparing the two sets of privacy laws (EU and the United States), it is clear that the majority of US laws governing the use of health information are still enacted at the federal or state level, such as the HIPAA law in the United States, which is closely linked to institutions that provide medical care, including hospitals, health insurance providers, and pharmacies. Many of these restrictions do not apply to other significant data obtained from social networks and other sources since they are outside of their purview. Furthermore, there are no explicit laws other than general guidelines from the US Trade Commission, regarding the fair information principles. The GDPR, in contrast, provides at least some limitations, although there are a number of important uncertainties over how to interpret the GDPR's approach to the regulation of social media data. The expansion of the definition of "health data" indicates that additional protections be applied to nonspecial categories of data as well as that the legal justifications for processing such types of data need to address this issue; requiring implicit consent could be one method. There is an urgent need to promote the creation of an alternate consent mechanism because the existing consent mechanisms cannot manage this additional strain.

In the literature, there has been debate over the topic of various data protection strategies. Cate et al., who oppose a more liberal approach, have argued in favor of user-driven self-regulation of data protection principles (Cate et al., 2013), which would put the burden on users. This viewpoint looks quite out of place and disregards the ethical obligations of healthcare companies toward their customers (Schneble et al., 2018).

13.5 IMPACT OF COVID-19 CONCERNING GDPR PRIVACY RULES

Regulators, like public and private entities, were not ready for the COVID-19 disaster. Many experts consider GDPR to be one of the strongest privacy laws in the world. The GDPR is not invalidated by the current crisis, according to the opinion of European regulators and the European data protection supervisor, its provisions are adaptable enough to accommodate the emergency measures while maintaining adequate safeguards. National governments, for instance, are allowed to act in the public interest under the GDPR, but they must restrict the data they use. Data privacy rules typically do not address the restrictions on free movement put in place by state authorities during the crisis. But these restrictions violate peoples' rights and liberties and might even be prohibited by other laws.

Employers are generally not allowed to collect health information from workers and visitors, as per the GDPR and other European regulations unless they are required to do so by law, whether to protect the interests of the workforce or the public or to adhere to employment law or other national laws. Data minimization, purpose limitation, openness, and data protection standards must all be closely adhered to when processing data, with certain exceptions. Only the minimum amount of information required for a given purpose may be collected by employers. They must tell those affected about data processing and guarantee that the data are protected properly. When the procedure is finished, it must be halted immediately and documented. Compared to the pre-COVID times, there is no change observed in this type of employee personal and health data collection (Daniel Mikkelsen et al., 2020).

The GDPR also restricts the ability to request employees' private phone numbers unless very necessary, such as when informing them of company policies or providing them with information about the COVID-19 outbreak. The GDPR limits the potential use of health-tracking technologies that certain nations have employed in response to COVID-19 by putting in place specific privacy and data protection protections. However, given that they adhere to the principles of the GDPR, national monitoring systems may be used, according to European data protection authorities. The GDPR forbids the use of more exact personal surveillance technologies, like those used in South Korea or China to stop the pandemic from spreading.

13.6 THE PERCEPTION OF PRIVACY IN DEVELOPING NATIONS

Privacy laws and technology are rarely taken into account in underdeveloped nations and humanitarian assistance efforts, where people are most at risk. It is a great matter of concern. In fact, privacy is frequently seen as a barrier to providing care for vulnerable people, even if bad privacy policies may make them even more vulnerable. Privacy is largely disregarded where it counts the most. Patient data are frequently vital in underdeveloped nations and humanitarian missions. Collection and sharing procedures are essential for public health surveillance. Information on patients must be accessible to several points of access and care providers in order to provide continuity of care. These details are necessary for research, process improvement, and life-saving procedures. The adoption of new technologies will further upend this already unstable environment now that eHealth is high on the agendas of governments, and the international community. Even more, information will be gathered and processed as a result of the use of technology like mobile devices and national data-processing centers. The continual effort to make previously independent health systems more interoperable made the dynamic of healthcare delivery even more difficult. Privacy need not be at odds with this, though. eHealth and privacy may even be mutually beneficial.

In the 1948 General Assembly of the United Nations, a statement was issued stating "Privacy is a human right!" (United Nations, 1948). Since then a lot of technological development took place extending privacy to online channels. Now, when we talk about privacy it widely covers offline as well as online portals of health exchange. The right to privacy is a reflection of digital channels strengthening the guidelines, directives for regulation, and adopted policies as established by international stakeholders like The Organization for Economic Cooperation and Development (OECD, 2002), Council of Europe (Council of Europe, 1985), Economic Community of West African States (ECOWAS, 2010).

The penetration of technology worldwide has been fast at an average rate of 58.4%. However, the same does not apply to developing countries like Africa where the rate remains to be an average low of 39.5%, leading to the global average rate of 62.4% (Internet World Stats, 2019). Elaborating further on African countries, the rate of individual countries varies from as low as 1.3% in Eritrea up to 89.9% in Kenya. Despite our strong belief that digital data are more secure than the paper trail of health information, Eric Santanen research in 2019 expressed otherwise. In the research titled "The value of protecting privacy" (Santanen, 2019), it was stated that "technology is so deeply embedded in daily lives, that people tend to lose awareness of their privacy while using these technologies." The study put another angle forward where the absence of technology can reduce the number of attacks on health information privacy. The debate over secure channels of information transfer in healthcare carries out on many fronts. Unequivocally, the penetration of technology is deep and it is here to stay. All work models are shifting to digital modes and those that are still in offline mode will soon be converted to online information. With this knowledge at hand, it becomes all the more important to look for secure channels to book health sessions and consult doctors. After all, privacy must not be compromised. Healthcare technology players like MediGence take great pride in delivering care to the masses with assured privacy and compliance with HIPAA and GDPR. As a global company, the services reach remote corners and make healthcare accessible to developing nations like Africa.

India is a fast-developing nation, but when it comes to privacy in healthcare, a lack of concern, among the public and well as the government, over the protection of patient data is very much on the edge. Sadly, The Indian Constitution has no recognition for the "Right to Privacy" to date. However, there is a constitutional right to "Personal Liberty" that can extend to include data privacy for individuals (Kumaraguru and Cranor, 2005). India is lagging behind with no Data Protection Bill like many other countries. However, a ray of hope at the beginning of 2019 came out of the darkness when the Government of India passed a "Personal Data Protection Bill." This bill is still in the pipeline and is expected to come into force in the near future (Datta, 2019). Although a latecomer, India is also working toward achieving universal health coverage for all under a National Health Policy mandate that was introduced in 2017 and is in implementation under the Ministry of Health and Family Welfare. The aim of this policy is to make healthcare accessible to all age groups and promote prevention as the primary approach to keep diseases at bay. This policy is referred to as Ayushman Bharat Digital Health Mission (ABDM) and will be included in the National Health Authority of India as an integral part. The ABDM will create unique health IDs for Indian citizens that will allow them to integrate and see all of their health data on a single platform. ABDM's primary objective also entails curating a repository of verified data for pharmacies, hospitals, clinics, doctors, and nurses with the goal to eliminate the chances of fraud in healthcare. This will not only bring down the cost of healthcare but also improve the quality of care and treatment around the country and put strict focus on the privacy of the healthcare data.

The low-income countries and especially low-income families within them are vulnerable to the misuse of data and need more protection and benefits from the state. Countries like Bangladesh also have problems encircling the privacy of citizens. Even though the constitution of Bangladesh safeguards communication privacy, the same is not extended to data privacy (Mishbah, 2019). The ICT Act was originally passed in 2006 for the protection of digital signatures and other electronic communications in the country but there were no provisions to safeguard data. To bridge this gap in data protection, a Digital Security Act was implemented in 2018 promoting topics like confidentiality and integrity with an aim to "protect the rights and privacy, security and economic interests of individuals in cyberspace." With this in effect, the future seems promising but not without its own set of ups and downs.

13.7 CHECKLIST TO ENSURE YOUR HEALTH DATA ARE SECURE

The typical medical practice of today looks very different from how it did only a few short years ago. Since many patients submit their medical histories online before visiting the office, there are fewer clipboard-attached forms that need to be filled out by patients in the waiting area. Instead of using paper charts in the exam room, doctors are typing their diagnoses into tablets. To give the best possible patient care, healthcare professionals are now collaborating their efforts digitally and also exchanging other data (Lehr, 2015).

These technical developments are among those enhancing contemporary healthcare, but they are also among those bringing new risks. Patients must step up for themselves and ask essential questions regarding their overall security if they hope to appropriately keep their sensitive healthcare data preserved. The healthcare organizations and providers must also put the responsibility on their shoulders to ensure their patient's data are safe and only authorized after completing due-diligence.

Here is a list of questions that every patient must ask their providers:

1. What is your organization doing to protect my health information on mobile devices?
2. How can you ensure/validate whether my data are secure?
3. Can I ask to delete/update my health data to the system in the future, if needed?
4. How do you adhere to privacy laws like HIPAA?
5. How do I share my data records/DICOM files/prescriptions with you without being concerned about them leaking out?
6. Does your organization use provisions like HIPAA, GDPR, nondisclosure agreements, encryption, and magic links to keep my data protected?

7. What is the legal framework for patients to sign with regards to their healthcare information, and how it will be treated in your organization?
8. Are you taking any measures to monitor and protect against insider threats?
9. Is your organization associated with third-party vendors to ensure risk assessment of my health data?
10. Will my data remain purely in the EHR systems or there will be a paper trail following it?

All above put in place, healthcare data security is crucial for providers to ensure and give confidence to the patients. Countries that rank on top as medical travel destinations are taking all measures to implement policies and take several technological initiatives to promise data sanity and security.

Your data only belongs to you and you should control and know who has its access. Next time, whether you consider traveling overseas for medical procedures or dealing with the physicians locally, do not hesitate to ask questions as stated in the checklist above.

KEYWORDS

- **privacy**
- **HIPAA**
- **GDPR**
- **healthcare documentation**
- **health data**

REFERENCES

American Health Information Management Association. The 10 Security Domains (Updated). *J. Am. Health Inf. Manage. Assoc.* **2012,** *83* (5), 50.

Cate, F. H.; Cullen, P.; Mayer-Schonberger, V. Data Protection Principles for the 21st Century. *Maurer Faculty*; Microsoft Corporation: Redmond, WA, 2013.

Council of Europe. *Convention for the Protection of Individuals with regard to Automatic Processing of Personal Data*, October 1, 1985.

Daniel Mikkelsen, S.; Soller, H.; Strandell-Jansson, M. *How Can European Companies Best Prevent Measures Intended to Control the COVID-19 Pandemic from also Undermining Data Privacy and Security?* June 15, 2020.

Datta, S. India Gears Up for Historic Data Protection Law. *Asia Times*, June 24, 2019. [Online]. Available: https://www.asiatimes.com/2019/06/article/india-gears-up-for-historic-data-protection-law/

Economic Community of West African States (ECOWAS). *Supplementary Act A/SA.1/01/10 on Personal Data Protection within ECOWAS*, February 16, 2010.

Gostin, L. Surveillance and Public Health Research: Personal Privacy and the "Right to Know." *Public Health Law: Power, Duty, Restraint*; University of California Press: Berkeley, CA, 2008.

HHS. Protecting Personal Health Information in Research: Understanding the HIPAA Privacy Rule, 2004c. http://privacyruleandresearch.nih.gov/pr_02.asp (accessed April 17, 2007).

Internet World Stats. *Africa Internet Users: Population and Facebook Statistics*, June 30, 2019. Available: https://www.internetworldstats.com/stats1.html

Kumaraguru, P.; Cranor, L. Privacy in India: Attitudes and Awareness. In *International Workshop on Privacy Enhancing Technologies*; Danezis, G., Martin, D., Eds.. Springer: Berlin and Heidelberg, 2005; vol 3856, pp 243–258.

Lehr, N. *Patient Data Security Questions Healthcare Companies Must Ask*, January 14, 2015.

Mishbah, H. Bangladesh Steps into the Data Protection Regime: What It Means for Us. *The Daily Star*, April 08, 2019 [Online]. Available: https://www.thedailystar.net/opinion/human-rights/news/bangladesh-steps-the-data-protection-regime-1726351

Odom-Wesley, B.; Brown, D.; Meyers, C. L. *Documentation for Medical Records*; American Health Information Management Association: Chicago, 2009; p 21.

Office of the National Coordinator for Health Information Technology. *Guide to Privacy and Security of Health Information*, 2012; p 5. http://www.healthit.gov/sites/default/files/pdf/privacy/privacy-and-security-guide.pdf (accessed August 10, 2012).

Organization for Economic Cooperation and Development (OECD). *OECD Guidelines on the Protection of Privacy and Transborder Flows of Personal Data*; OECD Publishing: Paris, 2002.

Pritts, J. *The Importance and Value of Protecting the Privacy of Health Information: Roles of HIPAA Privacy Rule and the Common Rule in Health Research*, 2008 (accessed March 15, 2008).

Rinehart-Thompson, L. A.; Harman, L. B. Privacy and Confidentiality. In *Ethical Challenges in the Management of Health Information*, 2nd ed. Harman, L. B., Ed.; Jones and Bartlett: Sudbury, MA, 2006; p 53.

Rognehaugh, R. *The Health Information Technology Dictionary*; Aspen: Gaithersburg, MD, 1999; p 125.

Santanen, E. The Value of Protecting Privacy. *Bus. Horizons* **2019,** *62* (1), 5–14.

Saver, R. Medical Research and Intangible Harm. *Univ. Cincinnati Law Rev.* **2006,** *74*, 941–1012.

Schneble, C. O.; Elger, B. S.; Shaw, D. The Cambridge Analytica Affair and Internet-Mediated Research. *EMBO Rep.* **2018,** *19* (8), e46579.

United Nations. *Universal Declaration of Human Rights: UDHR*, December 10, 1948.

CHAPTER 14

RELEVANCE OF POSTOPERATIVE CARE

"A treatment outcome solely depends on how good the recovery is."

ABSTRACT

Every surgical procedure demands the expertise of different physicians and specialists to help you go through the recovery phase smoothly. There is a huge gap in healthcare right now as most patients do not get readily available and right postoperative care plans for them; this makes them take wrong steps in their recovery period which can delay their healing or can leave a long-lasting after-surgery impact on their body. This chapter explores the role of telemedicine in delivering effective postoperative care, factors influencing a postoperative care plan, and rehabilitation after critical procedures. The chapter also talks about the importance of telerehabilitation in making care accessible to everyone globally.

The development of medical research has created new opportunities in the realm of surgery, where severe illnesses that once had deadly implications can now be treated using sophisticated surgical procedures. Despite surgical treatments becoming more popular among medical professionals, patients continue to view them as a risky method of addressing their particular ailments. Surgery does have an added risk of complications, but with a skilled surgeon in charge and the right care postsurgery, it is possible to optimize the advantages of surgery in aiding patients in recovering from a number of serious disorders. Following surgery, patients get care up until their release from the hospital or surgical facility, but in some situations, care is required even after being discharged from the hospital premises.

Medical, Dental, and Wellness Tourism: A Postpandemic Perspective. Mary Schreiber Swenson, PhD, and Amit Bansal (Authors).

That is where postoperative care takes the center stage. The goal of postoperative care is to immediately address the patient's physical and psychological requirements following surgery and help them gain back normal functions and lifestyle gradually. Each year, millions of patients have surgery throughout the world. Shorter hospital stays are a result of greater knowledge, improved anesthesia and surgical procedures, more awareness and knowledge about postoperative care parameters, and economic pressures (Bosch et al., 2002; Kehlet and Dahl, 2003).

For instance, it has been suggested that even elderly patients who undergo extensive colon surgery can be released from the hospital in two or three days (Basse et al., 2000). As a result of this pattern, patients and their families are now mostly responsible for postoperative recovery. In most cases, it looks like the possibility or capacity of patients to manage their own recovery was overlooked when making this big transition in the hospital stay phase post-treatment. This makes it all the more crucial to broaden the knowledge and comprehension of postoperative healing and put it into practice.

Postoperative care is a deciding factor in determining how well you will recover post a critical procedure and therefore it is strongly suggested by the experts to have a proper plan for rehabilitation after surgery before you are discharged home. Every surgical procedure demands the expertise of different physicians and specialists to help you go through the recovery phase smoothly. For instance, after a cardiac procedure, you may need the care of a physiotherapist, cardiologist, psychologist, and nutritionist but in the case of deep brain stimulation, you will need a neurologist in your care and rehabilitation plan. Every procedure is specific and demands the expertise of different experts. There is a huge gap in healthcare right now as most patients do not get the right postoperative care plans readily available for them. This makes them take wrong steps in their recovery period which can delay their healing or can leave a long-lasting after-surgery impact on their body. It is crucial to understand that healthcare should not just revolve around physical fitness but the mental trauma that may follow. Postsurgical depression is a common symptom of many critical procedures and must be dealt with under the care of the right physicians. Recovery is a very personal process and it differs from individual to individual hence, the rehabilitation also must be planned and curated after careful analysis of one's health needs by a professional. A right postoperative care plan helps in quicker and smoother recovery.

Major procedures including complex orthopedic procedures and spine or brain surgeries leave patients with restricted physical and mental capacities.

Through a variety of therapies, trained rehabilitation specialists work directly with patients to restore their physical capabilities, coordination, and balance. The essentials of right care after surgery depend on several factors including the type of procedure, complexity, age, and symptoms postsurgery. Conceptually, a patient needs hand-holding and continuous management not from the medical specialist only but the role of nutritionists, phycologists, and wellness experts cannot be overruled as they are equally the major contributors to a better recovery.

For instance, rehabilitation following surgery is quite helpful in helping patients restore their strength and reintegrate into daily life. It also plays a crucial part in managing pain and coordinating treatment in a proper and hygienic setting, which would not have been feasible otherwise.

Some of the benefits that patients get from postoperative care are the following:

- Helps in quickly restore full mobility
- Customized diet plan with the proper follow-up
- Helps in restoring control of the posture, balance, and coordination
- Aids in the control of pain
- Helps in reducing anxiety levels

14.1 ROLE OF TELEMEDICINE IN THE DELIVERY OF EFFECTIVE CARE AFTER A MEDICAL PROCEDURE

Preoperative and postoperative care are two areas of surgical treatment where telemedicine has been implemented recently (Gunter et al., 2016; Latifi et al., 2014). The use of telemedicine in postoperative care is of great interest because it has been shown to produce outstanding clinical results, improve patient satisfaction, increase accessibility while reducing wait times, and save both patients and healthcare systems money. The advancement of surgical care should concentrate on delivering the highest level of patient care with a focus on both patient and provider satisfaction while retaining the best possible access to care of experts that are required for proper long-term recovery.

Even though telemedicine has been widely embraced in many areas of surgical treatment, postoperative care is one area where it has attracted a lot of attention. Three methods of using telemedicine in the postoperative period have been employed most frequently: (1) scheduled follow-up (Hwa and Wren, 2013; Albert et al., 2013; Yoder et al., 2012), (2) routine monitoring

(Rao et al., 2012), and (3) management of emergent concerns as needed (Pirris et al., 2010; Sidana et al., 2014). Telemedicine has recently gained enormous popularity as a substitute for conventional medical appointments. Studies have shown that online video conference chats, which can take place from the patient's home, have the potential to replace routine follow-up visits to the clinic (Sharareh and Schwarzkopf, 2014). Surgical wound care evaluation services powered by smartphones and DICOM images have also attracted some interest (Segura-Sampedro et al., 2017).

Telemedicine delivers a lot of promise in the postoperative setting for improving the outcomes for patients, without which the healthcare intervention becomes fairly of limited use and leads to the absence of excellent results. Since telemedicine has been brought to use in the postoperative setting, many studies have reported better outcomes in recovery. The studies highlight that the patient recovery outcomes after a rehabilitation session, following the right diet plan, and care of other experts via an online platform are similar to the follow-up they receive in a traditional setting (Gunter et al., 2016). For instance, there are known greater incidence of complications with high-risk nonelective treatments, such as those involving orthopedic trauma but no additional health problems or complications have been reported following these operations when shifted to postoperative care using telemedicine as compared to the typical clinic follow-up settings (Sathiyakumar et al., 2015). It is important to note that no research has found that telemedicine follow-ups have statistically greater complication rates than conventional clinic follow-ups. Hence, it became a great option for patients during the COVID-19 pandemic and even after it ended.

Many researchers are now putting forward the debate on whether traditional follow-up visits are still necessary for low-risk surgeries given the potential that telemedicine has demonstrated in the postoperative situation. Earlier, for numerous minor and elective operations, doctors would typically schedule lengthy traditional office visits, as well as there would be subsequent follow-ups (Chen et al., 2014). Following low-risk elective surgeries, some practitioners have suggested that telemedicine follow-ups may be better, more efficient, and more convenient option. However, traditional follow-ups may still be essential for some patients, but in most cases, different parameters of care including sessions with therapists, nutrition, and psychologists can be delivered without any compromises with virtual mediums. Nevertheless, it seems that using telemedicine in the postoperative setting provides outstanding clinical outcomes across a wide range of surgical specialties that are equivalent to those following up in regular clinic visits.

14.2 FACTORS INFLUENCING THE POSTOPERATIVE PLAN

There are many factors that can impact the healing and recovery after a medical procedure. It is believed that factors such as prerecovery conditions, recovery time, encouragement and support, regular access to the right information, and setbacks throughout recovery might all have an impact on the course of recovery.

Preoperative circumstances of the patient, such as age, physical condition prior to surgery, diagnosis, and kind of operation, are considered to be significant predictors of recovery. It is thought that an individual's desire to participate in their recovery depends on their diagnosis and prognosis. According to this idea, severely ill patients were not subjected to the same standards and expectations as normal patients. Many reports suggest recovery varies depending on whether surgery was done as an elective or an urgent procedure. Additionally, it was observed that preoperative personal characteristics, such as prior surgical recovery experiences, could have either beneficial or unfavorable effects. Positive experiences were seen as enabling people to handle the difficulties of their subsequent interactions with healthcare more skillfully. At the same time, if prior bad surgical experiences did not return following the new surgery, the patients had positive mindsets which led to quicker recovery postsurgery.

Additionally, one of the most important factors for recovery postoperation was the time required. Many patients are discharged quickly from the hospital after a procedure. This can lead to improper management and care of their operated area during the healing time. They can later develop into lifetime problems. Not just that, individuals may feel trauma and stress after critical and ongoing procedures, which may or may not involve operating, like cancer. The time of recovery varies greatly on the type of disease a patient is suffering from and the methods employed to treat it. In cancer, chemotherapy is a common procedure that patients go through. The results of chemotherapy leave a long-lasting impact on the image and self-worth of the patient. Not just that, these kinds of treatments can often leave patients drained and in fear of recurring pain and trauma. An oncologist may treat the physical aspects of the condition but the psychological aspects need to be treated by a certified psychiatrist. During the late phases of recovery, diet also plays a crucial role. Patients who have undergone cardiac procedures may need a diet plan to avoid any future complications with their heart. They need a balanced and healthy meal plan which can only be guided and curated by a nutritionist. It is inconvenient for the patient to keep visiting physical clinics/hospitals after undergoing critical surgery and therefore, the ability to get in touch with a nutritionist, psychiatrist, and other experts via online modes have been bliss in the postoperative care segment.

14.3 REVIEW OF REHABILITATION POSTCRITICAL PROCEDURES

The research was conducted by the Agency of Healthcare Research and Quality (AHRQ) in which only 20% of the 1 million eligible patients who had a history of myocardial infarction, congestive heart failure, angioplasty, or heart surgery participated in cardiac rehabilitation (Rubin, 2019; World Health Organization, 2003). After the critical procedures, cardiac rehabilitation is thought of as the standard of therapy. Even among women, people of color, and low-income patients, adherence rates to rehabilitation and postoperative care are declining.

Despite the overwhelming evidence in favor of cardiac rehabilitation, the relationship between the cardiologist and other specialists still needs improvement. Right now, not every patient is completing even the minimum number of rehabilitation sessions that are mandated by standards (Rubin, 2019). And in many places, they are not being guided rightly or do not get the right package of care postsurgery. There would be a significant increase in lives saved if cardiac rehabilitation participation among eligible patients rose to 70%. About 180,000 fewer patients would require hospitalization (Ades et al., 2017).

Bariatric surgery is a significant treatment that may benefit from rehabilitation prior to or following surgery. Patients who engage in the proper workouts both before and after surgery reduce their risk of a cardiovascular event and see improvements in their anthropometric and physical fitness. In comparison to the previously described cardiac rehabilitation, the adherence rates in this discipline are significantly greater. Pouwels et al. demonstrated in their systematic review that the researchers considered satisfactory adherence levels in the nonsupervised and supervised exercise, which had a median of 57.3% (32.5–77.6%) and 64.5% (47.5–79.9%), respectively. No specific explanations were offered for this unique adherence rate (Pouwels et al., 2015).

People suffering from cancer are now living longer as there have been improvements in medical technology and treatment methods leading to early diagnosis and cutting-edge therapies. About 13.7 million cancer survivors in the United States made it through the year 2012 (Guy et al., 2013). The five-year survival rate for cancer patients in Iran is 66% for men and 72% for women (Rampisheh et al., 2014).

Despite the benefits, improved chances of survival might cause a number of physical and psychological problems as well as alterations in these patients' quality of life. Today, in addition to higher survival rates, treatment and care settings have altered as well, with patients typically continuing their care in outpatient facilities rather than hospitals (Sklenarova et al., 2015). As a result, patients are released from the hospital before fully recovering (Alaviani et al., 2015). Moving from the hospital to home is stressful for patients and their families, and if this transition is unsuccessful, it can have negative effects like rehospitalization, which puts a greater financial strain on patients, their families, and the healthcare system (Solan et al., 2015).

The fact that these patients no longer receive the supportive care and safety services they received from the hospital after being discharged distresses them. Postdischarge care is therefore crucial to the treatment of these individuals (Harrison et al., 2012) Patient-centered care (Harrison et al., 2009), which is care based on the patients' needs, is one of the quality care measures for these patients (Epstein and Street, 2011). Determining the postdischarge care requirements of cancer patients is, therefore, necessary in order to provide them with high-quality postdischarge care. There is evidence to support the idea that anticipating requirements and making plans accordingly can lower medical expenses while also boosting patient independence (Baghaei et al., 2013).

Additionally, failing to meet a patient's postdischarge demands results in several issues and failure to cater to them. Evidence reveals that some of these issues include elevated anxiety, deterioration of the patients' mental and physical health, emotional issues, and a decreased capacity for adaptation (Rampisheh et al., 2014). Cancer exposure results in pain, stress, and a great deal of pressure on the patient and family, which will increase after discharge. Therefore, for these patients and their families, "protection from the stress and excessive pressure" is a crucial need. After discharge, patients and their families may still have access to the treatment team or knowledgeable experts to help them with postoperative care plans. A study found that while most patients feel abandoned after discharge, they still need to maintain regular touch with the necessary healthcare experts (Mikkelsen et al., 2008).

In recent years, neurorehabilitation has seen many changes. Most patients with common neurological conditions like stroke, Parkinson's disease, spinal cord injury, severe brain injury, spasticity, and cognitive difficulties who needed traditional neurorehabilitation techniques to manage a healthy and long life are now resorting to modern ways. The efficiency of rehabilitation strategies has reportedly been improved by new technologies. When the pandemic took the nations by storm, the need to shift management and care for neuroconditions to online platforms became a necessity rather than a choice. This is when the need for experts like Speech Language Therapists/ Occupational therapists increased and telerehabilitation for neurological conditions emerged as a life-saver.

Since then, there have been major changes in rehabilitation and enablement services as a result of changing demographics and an increase in community care. There is growing evidence that rehabilitation programs improve independence and well-being as people live longer and have more complex health demands (Wade, 2020). Neurorehabilitation services joined the response after the severe acute respiratory syndrome coronavirus 2 outbreak with the aim to help with the global health emergency and ensure that supply meets the demand of patients in the crucial time (World Health Organization, 2020). Speech and language therapy (SLT), which assesses and treats persons with speech, language, and communication disorders, is a crucial component of the multidisciplinary approach to neurorehabilitation. SLT experts, psychiatrists, and occupational therapists, all joined hands and became available on virtual platforms to dispatch care to patients who needed it desperately.

The video conferencing made it possible for patients and their loved ones to manage their health rightly and also gave an alternative to traditional hospital care visits. The long-term care plan to deal with neuroconditions via virtual channels slowly but gradually is making care available and accessible all around the world. The access, comfort, and care brought about in the care with personalized and well-curated recovery plans make it wholesome. It's a breakthrough and no patient is ever going back to the old ways completely.

Some studies suggest that up to 50% of individuals who are advised to receive any type of rehabilitation do not (Argent et al., 2018) when taken as a whole. Therefore, noncompliance could eventually lead to unresolved long-term health issues. In general, nonadherence can put more of a financial strain on the healthcare system than following a rightly curated postoperative care plan (Bassett et al., 2007). That is why it is crucial to have the right care after treatment so that patients can lead healthy and longer lives.

KEYWORDS

- **telerehabilitation**
- **postoperative care**
- **recovery**

REFERENCES

Ades, P. A.; Keteyian, S. J.; Wright, J. S.; et al. Increasing Cardiac Rehabilitation Participation from 20% to 70%: A Road Map from the Million Hearts Cardiac Rehabilitation Collaborative. *Mayo Clin. Proc.* **2017,** *92*, 234–242.

Alaviani, M.; Khosravan, S.; Hoseyni, R.; Hoseyni, M. Caring Needs of Discharged Patients from Medical-Surgical Wards of Gonabad Hospitals. *Nurs. J. Vulner.* **2015,** *2*, 25–35.

Albert, M. V.; McCarthy, C.; Valentin, J.; et al. Monitoring Functional Capability of Individuals with Lower Limb Amputations Using Mobile Phones. *PLoS One* **2013,** *8*, e65340. doi:10.1371/journal.pone.0065340.

Argent, R.; Daly, A.; Caulfield, B. Patient Involvement with Home-Based Exercise Programs: Can Connected Health Interventions Influence Adherence? *JMIR Mhealth Uhealth* **2018,** *6*, e47.

Baghaei, R.; Sharifi, M.; Mohammadpour, Y.; Sheykhi, N. Evaluation of the Effects of Educational Package on Controlling the Complications of Chemotherapeutic Agents on Symptom Scales of Quality of Life in Patients with Breast Cancer Undergoing Chemotherapy. *J. Urmia Nurs. Midwifery Fac.* **2013,** *11*, 667–679.

Basse, L.; Hjort Jakobsen, D.; Billesbolle, P.; Werner, M.; Kehlet, H. A Clinical Pathway to Accelerate Recovery after Colonic Resection. *Ann. Surg.* **2000,** *232* (1), 51–57.

Bassett, S. F.; Prapavessis, H. Home-Based Physical Therapy Intervention with Adherence-Enhancing Strategies Versus Clinic-Based Management for Patients with Ankle Sprains. *Phys. Ther.* **2007,** *87*, 1132–1143.

Bosch, F.; Wehrman, U.; Saeger, H. D.; Kirch, W. Laparoscopic or Open Conventional Cholecystectomy: Clinical and Economic Considerations. *Eur. J. Surg.* **2002,** *168* (5), 270–277.

Chen, D. W.; Davis, R. W.; Balentine, C. J.; et al. Utility of Routine Postoperative Visit after Appendectomy and Cholecystectomy with Evaluation of Mobile Technology Access in an Urban Safety Net Population. *J. Surg. Res.* **2014,** *190*, 478–483. doi:10.1016/j.jss.2014.04.028.

Epstein, R. M.; Street, R. L., Jr. The Values and Value of Patient-Centered Care. *Ann. Fam. Med.* **2011,** *9*, 100–103.

Gunter, R. L.; Chouinard, S.; Fernandes-Taylor, S.; et al. Current Use of Telemedicine for Post-discharge Surgical Care: A Systematic Review. *J. Am. Coll. Surg.* **2016,** *222*, 915–927. doi:10.1016/j.jamcollsurg.2016.01.062.

Guy, G. P., Jr.; Ekwueme, D. U.; Yabroff, K. R.; Dowling, E. C.; Li, C.; Rodriguez, J. L.; et al. Economic Burden of Cancer Survivorship among Adults in the United States. *J. Clin. Oncol.* **2013,** *31*, 3749–3757.

Harrison, J. D.; Young, J. M.; Price, M. A.; Butow, P. N.; Solomon, M. J. What Are the Unmet Supportive Care Needs of People with Cancer? A Systematic Review. *Support Care Cancer* **2009,** *17*, 1117–1128.

Harrison, S. E.; Watson, E. K.; Ward, A. M.; Khan, N. F.; Turner, D.; Adams, E.; et al. Cancer Survivors' Experiences of Discharge from Hospital Follow-Up. *Eur. J. Cancer Care (Engl.)* **2012,** *21*, 390–397.

Hwa, K.; Wren, S. M. Telehealth Follow-Up in Lieu of Postoperative Clinic Visit for Ambulatory Surgery: Results of a Pilot Program. *JAMA Surg.* **2013,** *148*, 823–827. doi:10.1001/jamasurg.2013.2672.

Kehlet, H.; Dahl, J. B. Anaesthesia, Surgery and Challenges in Postoperative Recovery. *Lancet* **2003,** *362* (9399), 1921–1928.

Latifi, R.; Mora, F.; Bekteshi, F.; et al. Preoperative Telemedicine Evaluation of Surgical Mission Patients: Should We Use It Routinely? *Bull. Am. Coll. Surg.* **2014,** *99*, 17–23.

Mikkelsen, T. H.; Søndergaard, J.; Jensen, A. B.; Olesen, F. Cancer Rehabilitation: Psychosocial Rehabilitation Needs after Discharge from Hospital? *Scand. J. Prim. Health Care* **2008,** *26*, 216–221.

Pirris, S. M.; Monaco, E. A., 3rd.; Tyler-Kabara, E. C. Telemedicine through the Use of Digital Cell Phone Technology in Pediatric Neurosurgery: A Case Series. *Neurosurgery* **2010,** *66*, 999–1004. doi:10.1227/01.NEU.0000368443.43565.2A.

Pouwels, S.; Wit, M.; Teijink, J. A.; Nienhuijs, S. W. Aspects of Exercise before or after Bariatric Surgery: A Systematic Review. *Obes. Facts* **2015,** *8*, 132–146.

Rampisheh, Z.; Motamed, N.; Amiri, M.; Ostevar, A. Survival Analysis of Cancer Patients Registered in the Cancer Registry System of Bushehr Province 2001–2013. *PhD Thesis*; Bushehr University of Medical Sciences and Health Services: Iran, 2014.

Rao, R.; Shukla, B. M.; Saint-Cyr, M.; et al. Take Two and Text Me in the Morning: Optimizing Clinical Time with a Short Messaging System. *Plast. Reconstr. Surg.* **2012,** *130*, 44–49. doi:10.1097/PRS.0b013e3182547d63.

Rubin, R. Although Cardiac Rehab Saves Lives, Few Eligible Patients Take Part. *JAMA* **2019,** *322*, 386–388.

Sathiyakumar, V.; Apfeld, J. C.; Obremskey, W. T.; et al. Prospective Randomized Controlled Trial Using Telemedicine for Follow-Ups in an Orthopedic Trauma Population: A Pilot Study. *J. Orthop. Trauma* **2015,** *29*, e139–e145. doi:10.1097/BOT.0000000000000189.

Segura-Sampedro, J. J.; Rivero-Belenchon, I.; Pino-Diaz, V.; et al. Feasibility and Safety of Surgical Wound Remote Follow-Up by Smart Phone in Appendectomy: A Pilot Study. *Ann. Med. Surg. (Lond.)* **2017,** *21*, 58–62. doi:10.1016/j.amsu.2017.07.040.

Sharareh, B.; Schwarzkopf, R. Effectiveness of Telemedical Applications in Postoperative Follow-Up after Total Joint Arthroplasty. *J. Arthroplasty* **2014,** *29*, 918–922.e1. doi:10.1016/j.arth.2013.09.019.

Sidana, A.; Noori, S.; Patil, N. Utility of Smartphone Camera in Patient Management in Urology. *Can. J. Urol.* **2014,** *21*, 7449–7453.

Sklenarova, H.; Krümpelmann, A.; Haun, M. W.; Friederich, H. C.; Huber, J.; Thomas, M.; et al. When Do We Need to Care about the Caregiver? Supportive Care Needs, Anxiety, and Depression among Informal Caregivers of Patients with Cancer and Cancer Survivors. *Cancer* **2015,** *121*, 1513–1519.

Solan, L. G.; Beck, A. F.; Brunswick, S. A.; Sauers, H. S.; Wade-Murphy, S.; Simmons, J. M.; et al. The Family Perspective on Hospital to Home Transitions: A Qualitative Study. *Pediatrics* **2015,** *136*, e1539–e1549.

Wade, D. T. What Is Rehabilitation? An Empirical Investigation Leading to an Evidence-Based Description. *Clin. Rehabil.* **2020,** *34*, 571–583. doi:10.1177/0269215520905112.

World Health Organization. *Adherence to Long Term Therapies—Evidence for Action*; World Health Organization: Geneva, Switzerland, 2003.

World Health Organization. *Rolling Updates on Coronavirus Disease (COVID-19)*; WHO: Geneva, 2020. Available online at: https://www.who.int/emergencies/diseases/novel-coronavirus-2019/events-as-they-happen (accessed November 13, 2020).

Yoder, L. H.; McFall, D. C.; Cancio, L. C. Use of the Videophone to Collect Quality of Life Data from Burn Patients. *Int. J. Burns Trauma* **2012,** *2*, 135–144.

CHAPTER 15

HOW TECHNOLOGY IS CONTINUOUSLY AND TREMENDOUSLY IMPROVING MEDICAL TRAVEL AS AN INDUSTRY

ABSTRACT

The size of the global medical tourism industry is gradually increasing year on year, and it is anticipated to increase at a CAGR of 32.51% from 2022 to 2030. Some of the key factors driving the market include cost savings and extra benefits to visitors, including enhanced healthcare, cutting-edge technologies, ground-breaking pharmaceuticals, sophisticated gadgets, superior hospitality, and personalized care. Technology has made it easy for patients to locate the right country and right physicians for the services and treatments they want to get. Additionally, it is now possible to compare costs, hospitals, and doctors by just going over health websites on the internet. This chapter explores the role of technology in improving the medical travel industry, various parameters involved in this rapid expansion, and increasing concerns about health data privacy.

In the past decade or so, there has been tremendous development and movement in the medical industry, merging healthcare with travel sectors forming one integrated whole. The industry has always been robust with many tightly wounded and integrated sectors like wellness and information, communication and technology, and travel and tourism. This has led to a phenomenal improvement and advancement in mobility worldwide for availing the best quality treatment. When we talk of healthcare travel, the history dates back to ancient times when people used to cross geographical boundaries in search of the right treatment and care services.

Medical, Dental, and Wellness Tourism: A Postpandemic Perspective. Mary Schreiber Swenson, PhD, and Amit Bansal (Authors).

Ancient cultures acknowledged the healing benefits of thermal medicine, hot springs, and holy temple baths (Gianfaldoni et al., 2017). Around 4000 BC, the Sumerians constructed the oldest health facilities near hot springs, while the ancient Greeks built the Asclepias temple in honor of the God of Medicine (Health-Tourism.com, 2020). This became one of the world's first travel destinations where people from all over the world started to travel to get treatment and solutions for their health issues. This way, Greeks were the first ones to lay the foundation of a comprehensive traveling network in the healthcare domain, and ever since it has evolved and flourished.

Modern healthcare travel has expanded to incorporate leisure activities and well-being improvement because patients now need time to rest and recover. Destination marketers are now required by evolution to reconsider why healthcare travelers choose to visit a particular location. The term "health tourism" was eventually coined by fusing the ideas of travel, tourism, and wellness. Since the 17th century, the phrase "health tourism" or "medical tourism" has been in usage (Yusof et al., 2019). Although the terms "health tourism," "medical tourism," and "wellness tourism" are sometimes used interchangeably, it may be essential to clarify that they are all actually distinct concepts (Wong and Musa, 2013). Therefore, in this context, the term "health tourism" refers to travelers who go elsewhere for a variety of reasons to obtain health, medical, and/or wellness services.

We live in the era of innovation, there are numerous newly discovered diseases and so are new treatment options and advancements in healthcare. These treatments may not always be available in your country and this might be one of the reasons for considering traveling overseas for the right treatment.

Healthcare has always evolved progressively alongside medical technology (Allen, 2019; Lobo, 2020). Thus, the development of technology is redefining the need for healthcare travel by improving medical accuracy and quality while also delivering healthcare services closer and faster to the patients. The demand for advanced and cultured medical procedures and equipment superior to individualized ones has progressively been incorporated into the growth of health tourism (Bhattacharyya, 2020; Hong, 2016; Medical Technology, 2020).

In today's technology-driven economy, the term "Industrial Revolution 4.0" (IR 4.0) is frequently used. In contrast to Industry 1, which introduced the use of steam-powered machinery to boost output, Industry 2 popularized the use of electricity and mass production techniques. The third Industrial Revolution (IR) delivered the first wave of technological digitization, enabling economies to expand quicker and beyond political and geographical boundaries. According to Bernasconi (2016), the fourth IR, which is an evolution of Industry 3.0 that connects the physical, digital, and biological domains, is still upending traditional economic models and social norms. Internet of Things, robotics, virtual reality, artificial intelligence, cloud computing, and big data analytics are a few examples of IR 4.0 technology.

Despite technological developments, both the general tourism industry and the development of the health tourism sector gained. With a significant revolution within the mobility enhancement concept, IR 4.0 is taking the healthcare business by storm (Landman, 2018). With the help of mobile healthcare, it is anticipated that healthcare would become more accessible, economical, and secure. Telemedicine, wearables, digital sensors, biotelemetry, remote patient monitoring, virtual rehabilitation, and intelligent fabric are a few of the mobile healthcare infrastructure and technology that will contribute to creating the future.

It has helped a lot of patients, but it has also made it easier for healthcare professionals to organize their work, gather data, and give real-time updates. The most recent advancements in healthcare technology give a significant competitive advantage, especially when it comes to creating quicker and more efficient treatments (Bernasconi, 2016).

One of the first stages of IT-enabled collaboration in the healthcare sector is telemedicine. Healthcare experts have largely recognized the advantages of telemedicine implementation, notably in the treatment of chronic diseases, especially after the COVID-19 pandemic. Patient who seeks to get treatment from a specialist outside their reach can do so now. It increases patients' virtual mobility, reducing the influence of a patient's physical location when seeking medical care. This virtual reachability is a great trust-builder that later promotes patients to travel for treatment to the same doctor in their practicing country boosting the number of medical travel. This makes it possible for folks in rural locations to access the best medical care with help of technology.

Over 80% of the patients in a study at the Mayo Clinic in the United States expressed interest in utilizing mobile health applications, demonstrating an astounding level of acceptability for mobile technology. Health services were identified by the US Secretary of Health and Human Services as one of the most important health-related technologies in 2011. (Nehra et al., 2017). Bringing health services online makes the process of booking medical travel easy. However, with many agents in between, the process can sometimes get confusing. This is why healthcare platforms like MediGence are delivering healthcare services with complete transparency and many complimentary benefits. The idea is to make a patient's travel journey smooth and effective so that they can recover without stressing about where to stay, how to manage their meals, and other such factors.

When it comes to improving pre-and postoperative care from a medical and customer service viewpoint, technology is especially relevant and helpful for health tourism. Anesthesiologists and surgeons can conduct teleconsultations from afar to evaluate how well foreign visitors to medical tourism are recovering from their wounds and recuperating. With the aid of teleconferencing and/or telehealth apps, medical staff at the health tourism destination can keep in touch with the patients' primary care doctors back home. This is increasingly becoming a great motivator and giving more and more individuals the confidence that they will be getting the right care even when they are back in their hometown.

15.1 THE GROWTH OF THE MEDICAL TOURISM INDUSTRY

The size of the global medical tourism industry is gradually increasing year on year, and it is anticipated to increase at a CAGR of 32.51% from 2022 to 2030. Some of the key factors driving the market include cost savings and extra benefits to visitors, including enhanced healthcare, cutting-edge technologies, ground-breaking pharmaceuticals, sophisticated gadgets, superior hospitality, and personalized care.

The market is anticipated to grow rapidly over the course of the forecast period. The main driver of the market's expansion is either the exorbitant cost of the treatment, long waiting, lack of quality, or nonavailability of the medical treatment in the country. The sector is also being fueled by an increase in demand for services like gender reassignment surgery, reproductive therapy, dental reconstruction, and cosmetic surgery that are not covered by insurance. Technology has made it easy for patients to locate the right country and right physicians for the services and treatments they want to get.

Additionally, it is now possible to compare costs, hospitals, and doctors by just going over health websites on the internet. For instance, an angioplasty surgery in the United States costs an individual about USD 32,000 to USD 37,000 compared to approximately USD 8000 to USD 8500 in Malaysia. People fly to Thailand, Singapore, India, and Malaysia for medical care from the United States and Europe, saving a lot on treatment costs. Malaysia's medical treatment costs are 20–40% less expensive. In these nations, a sizable number of foreigners seek cosmetic treatments like liposuction, dental implants, breast implants, and facelifts. The expansion of the global market will be fueled by improvements in the healthcare infrastructure and the accessibility of high-quality healthcare services at reasonable costs. Without complete transparency and knowledge about the costing structure, it would have been impossible for the patients to make the decision about traveling for treatment.

Medical travel has been one of the major industries helping top medical destinations (countries) in boosting their foreign reserves that eventually contributes in the overall GDP of the nation. Many countries have been taking several initiatives in creating and promoting medical tourism. For instance, to draw patients from North America, Latin American nations like Mexico, Brazil, and Colombia have made significant investments in their healthcare systems. A few Eastern European nations, like Poland, Hungary, and the Czech Republic, have also made names for themselves as regional medical hubs, drawing patients from Western Europe. The European Union passed Directive 2011/24/EU on patients' rights in cross-border healthcare in March 2011 in response to the rise of medical tourists in Europe, which established the rights of EU residents to seek medical treatment in member nations (European Commission, 2016). Israel, Turkey, and the United Arab Emirates have deliberately built their healthcare systems to draw patients from Europe and other nearby nations in the Middle East (for instance Kannan, 2014).

15.2 UNDERSTANDING THE PARAMETERS PROMOTING EXPANSION OF THE MEDICAL TRAVEL INDUSTRY

New consumer-focused tools are being offered in the travel and tourism sectors. The effects of technology are highlighted by research in the tourism sector, which also aids in the creation of new strategies for enhancing the happiness of medical tourists. For instance, information technology has significantly facilitated easy access to information on the procedures and treatments that hospitals around the world may offer for international patients (Levebvre and Bochatan, 2002). Additionally, there are a number of information technologies that can be successfully implemented in the medical tourism industry, such as point-of-sale systems for cosmetic services, electronic health records systems, and websites of destination management organizations (Qi et al., 2008).

Medical tourism is the planned, ongoing maintenance and restoration of one's physical and emotional health, according to Carrera and Bridges (2006). Therefore, the main factor influencing medical tourists' travel choices is their need for quality medical treatments. More generally, the final choices chosen by aspirational medical tourists are impacted by their web research for dependable, secure, fast, and affordable healthcare options and services.

The main factors influencing the quest for alternative medical treatment and healthcare options are cost, accessibility, and availability. Public healthcare frequently provides necessary treatments in many nations, especially those in developed regions of the world, but private medical insurance coverage is expensive and may penalize people with preexisting medical issues. Longer wait times for interventions in public healthcare, which can have a negative influence on patients' health and wellbeing, are frequently observed (Patrick and Puterman, 2008). Peer pressure, media exposure, and worldwide ideas on idealized appearance are factors that influence women's decision to have cosmetic surgery, while societal acceptability of the procedure acts as an enabler (Sharp et al., 2014). All these factors have an impact on the decision-making of an individual for which they might later resort to travel to other nations. The end goal is to find the right care with the expertise of skilled and certified doctors. And with help of technology penetrating every part of the world, individuals now have access to an endless source of data to match, check, and make a worthy decision.

Patients who are suffering from critical conditions like some kind of an "acquired defect" brought on by cancer, burns, or other sorts of trauma also frequently resort to travel. However, the cases of travel for treatment fell considerably during the COVID-19 pandemic, but since then the rise in numbers is clearly visible. And rightfully so, patients prefer seeking treatment from a location where they will be assured of getting only the best care within their budget.

15.3 AFFORDABILITY

Medical tourists weigh the cost of foreign travel, the favorableness of currency exchange rates, and the rising standard of medical care in some developing nations (Swain and Sahu, 2008). The affordability of medical treatment has turned into a crucial factor in changing the patterns and flows of medical travelers as the technological divide has closed globally. A number of medical destinations also adapt to the changing demands of the "tourist-patient" for inexpensive medical care, giving them a financial advantage and a competitive edge in the worldwide market. As an illustration, spine surgery in the US = $30,000–$55,000. However, the same medical procedure in Thailand costs $18,000–$25,000. The vastly reduced costs apply to an expanding range of healthcare services offered throughout Central Europe and Asia (Bostan et al., 2016).

15.4 ACCESSIBILITY AND AVAILABILITY

Accessibility and availability of particular healthcare services (as well as wait times) in patients' home countries are two significant factors that influence medical tourism. More specifically, the need for medical travel is brought on by the lack of temporary unavailability of the necessary medical services in the patients' home nation. Additionally, if treatment procedures are postponed owing to lengthy wait times, priority waiting lists, or a lack of available organs for transplant surgeries, medical tourism may be a possibility. Due to the long waits for medical consultations and interventions, Dawn and Pal (2011) point out that the majority of medical tourists from developed nations like the United Kingdom, Japan, the United States, and Canada seek their treatments abroad. Similar to this, medical tourism is fueled by the ability of the Japanese healthcare system to meet the medical needs of the nation's rapidly expanding older population. For annual medical exams, for instance, many Japanese companies transfer their staff to nearby nations. This type of medical travel highlights the value of organ transplant travel, which saves thousands of lives annually.

15.5 EASE OF CHOICE

In the medical tourism sector, traditional travel agents continue to hold a strong position, which can encourage opportunistic conduct. Instead of developing strong relationships between medical institutions and their patients, the involvement of intermediaries in medical tourism has resulted in a paternalistic form of partnerships (Lunt et al., 2010). The medical tourism packages provided by these organizations are frequently marked up significantly from the actual costs of the healthcare service providers and lack transparency about the quality of treatment and other factors. Additionally, as these third-party agents have limited connections to particular healthcare providers to whom they refer all of their consumers, there is always a gap in the care that can be availed otherwise (Herrick, 2007). Since they are experts in a certain niche or service, intermediaries may steer people toward unsuitable doctors who charge exorbitant referral fees and add-on markups. Worse yet, because they are not actual healthcare providers, these medical tourism intermediates are unlikely to be held legally liable for any mishaps (Connell, 2011).

Not to mention, not all intermediaries deliver a worse experience. Some healthcare companies have stepped up in the field to change the scenario. To eliminate this long prevailed issue, certain healthcare-focused and patient-centric platforms have come up in the digital space such as MediGence. The key motive of such providers is to enable individuals to access the right care with transparency and many additional perks that are very much necessary when choosing to travel for treatment to a foreign country.

Additionally, with the assistance of technology, patients can interact directly with healthcare professionals. They can so objectively confirm their credentials, certifications, and associated charges. Removing the needless stress and anxiety of having to plan every part of their medical travel and tourism on their own can help establish confidence and trust. The legitimacy of the industry will also increase with the streamlining of the medical tourism infrastructure by eliminating or curtailing the influence of nonvalue-adding intermediaries (Raman and Varshney, 2017). As a result, the technology has the ability to create new value since it lessens information asymmetry by making information and knowledge about medical facilities more transparent. As a result, technology can help with a variety of problems and provide patients the chance to consider their options and advantages (Ehrbeck et al., 2008).

15.6 ENSURING TRUST AND TRANSPARENCY

According to Sirdeshmuhk et al. (2002), trust is the belief that a service provider is trustworthy and capable of keeping its commitments. This means that the healthcare provider is skilled, that their medical staff is knowledgeable, and that the overall medical facility is very competent at enhancing the health of its international patients. More people are now aware of healthcare services with the abundance of information on medical websites. The usage of ICTs (information and communication technologies) aids the patient-tourist in decision-making and might help them choose the right medical packages within their budget. Through innovative marketing avenues, media attention, and social media sites like Facebook, Twitter, and YouTube, medical tourism businesses work to draw clients. ICTs are crucial for enabling medical travel, particularly during the first stages of information research and facility selection. The largest problem facing medical tourism providers, according to Connell (2006), is convincing potential clients of their qualifications, the caliber of their pre- and postprocedure care, and general patient safety.

In order to reduce information asymmetry between themselves and potential patients, medical facilities use a variety of online and offline platforms to signal their qualifications. Although they are signal receivers, potential patients lack the knowledge and resources to confirm these "unobservable" credentials from the medical center (Connelly et al., 2011). A potential patient will not accept the medical providers' unobservable characteristics, such as the standard of care and the expertise of specialists if qualifications cannot be verified. Technology can be utilized to connect the numerous certification bodies and authorities to the qualifications claims made by healthcare providers. In this sense, technology makes it easier to verify credentials, such as the accreditations and training of healthcare providers, doctors, and specialists, as well as the verification of online reviews.

15.7 INCREASING CONCERNS ABOUT PRIVACY

Patient privacy is a key element of medical tourism. Patients who travel for medical care frequently entrust their private medical information to unreliable healthcare professionals or middlemen (Connell, 2006). According to a poll of American medical tourists, the second most important factor they took into account after treatment costs were "privacy and secrecy of treatment" (Singh, 2013). In a study on wellness tourism in the Asia-Pacific region, privacy was shown to be an important consideration for higher paying visitors (Kucukusta and Guillet, 2014). Regardless of how far away their medical tourism destination is from them, some tourist patients feel more secure knowing that their privacy will be maintained when it comes to the sensitive nature of their medical records.

By adhering to international standards and certifications, this issue with secrecy and medical tourists' expectations can be resolved. Platforms that adhere to HIPAA and GDPR regulations ensure that patients' personal information is secure and that no data are utilized without their consent. The ability of a foreign government to uphold adequate privacy laws and secure sensitive medical data may raise concerns in the minds of some potential medical tourists. This mistrust is dispelled when patients are assured that they will receive medical care from qualified and licensed doctors in hospitals that have been JCI-certified or nationally board-certified. Patients tend to feel more at peace when they are fully informed and transparent about how their personal information is used. As a result of their long-lasting control, medical tourists' privacy will not be an issue even after having treatment.

15.8 IMPACTS OF COVID-19 ON GLOBAL TOURISM AND THE ROAD AHEAD

It is crucial to remember that global tourism has already been subject to a variety of problems. Major disruptive events that occurred between 2000 and 2015 include the September 11 terrorist attacks (2001), the SARS pandemic in 2003, the global economic crisis that developed in 2008/2009, and the Middle East Respiratory Syndrome outbreak in 2015. Only SARS (−0.4%) and the global economic crisis (−4.0%) resulted in decreases in international arrivals, with none of them causing a longer term reduction in the worldwide development of tourism. This would imply that the tourism industry as a whole has been resilient to outside shocks. There is enough data, though, to suggest that the COVID-19 pandemic will have extraordinary effects and recovery times.

Travel restrictions began to appear in most nations by the end of March when COVID-19 instances surged and spread throughout the globe (local lockdown started on January 23 in the Wuhan region). Many nations closed their borders to the movement of noncitizens and nonresidents. While some countries had partial border closures, such as limitations on people arriving from specific other nations or countries where not all types of borders are closed (air, land, and sea). Almost all other nations implemented at least some travel restrictions, such as travel bans from specific countries, arrival quarantines, and/or health certificate requirements, in response to this unprecedented response, which resulted in the closing of borders in a wide range of industrialized countries to all foreign nationals. The impact was huge on medical travel. People who were in need of treatments that required traveling outside their geographies had to resort to alternatives provided by technology. Telemedicine and facilities like virtual second opinion and rehabilitation played a crucial role in health management during this time.

Once the restrictions began lifting up, patients who wanted to see their virtual care providers in person started moving quickly. There was concern among researchers and experts if the travel market in healthcare will ever bloom again and to everyone's surprise, as soon as restrictions began lifting up, people started falling back to their normal habits. Although technology took over a big chunk of the health space, now people read reviews and testimonials and go through a list of providers from different locations, compare packages online and then travel for their treatment but overall the sector is booming and will keep rising post-COVID-19 in the coming years.

KEYWORDS

- **healthcare technology**
- **medical travel**
- **privacy**
- **medical tourism**
- **travel for treatment**
- **health data**
- **privacy**

REFERENCES

Allen, S. *2019 Global Health Care Outlook*, 2019. Available at: www2.deloitte.com/global/en/pages/lifesciences-and-healthcare/articles/global-health-care-sector-outlook.html (accessed October 31, 2019).

Bernasconi, S. *MedTech Views*, 2016. Available at: www.medtechviews.eu/article/role-medicaltechnology-4th-industrial-revolution (accessed October 31, 2019).

Bhattacharyya, S. M. Medical Tourism: Role of Telemedicine, 2020. Available at: www.asianhhm.com/ healthcare-management/medical-tourism-telemedicine (accessed April 13, 2020).

Bostan, I.; Roman, T.; Lazar, C.; Manolica, A. The Current Trends and Opportunities in the Industry of Medical Tourism. *Rev. Tur.—Studii Cercet. Tur.* **2016,** *22*, 58–63.

Carrera, P. M.; Bridges, J. F. Globalization and Healthcare: Understanding Health and Medical Tourism. *Expert Rev. Pharmacoecon. Outcomes Res.* **2006,** *6* (4), 447–454.

Connell, J. *Medical Tourism*; Wallingford: Cabi, 2011.

Connelly, B. L.; Certo, S. T.; Ireland, R. D.; Reutzel, C. R. Signaling Theory: A Review and Assessment. *J. Manage.* **2011,** *37* (1), 39–67.

Dawn, S. K.; Pal, S. Medical Tourism in India: Issues, Opportunities and Designing Strategies for Growth and Development. *Int. J. Multidiscipl. Res.* **2011,** *1* (3), 7–10.

Ehrbeck, T.; Guevara, C.; Mango, P. D. Mapping the Market for Medical Travel. *McKinsey Q.* **2008,** *11*.

European Commission. *Cross-Border Care*, 2016. http://ec.europa.eu/health/cross_border_care/policy/index_en.htm (accessed November 12, 2016).

Gianfaldoni, S.; Tchernev, G.; Wollina, U.; Roccia, M. G.; Fioranelli, M.; Gianfaldoni, R.; Lotti, T. History of the Baths and Thermal Medicine. *Maced. J. Med. Sci.* **2017,** *5* (4), 566–568. doi:10.3889/oamjms.2017.126.

Grand View Research.com. *Medical Tourism Market Size, Share & Trends Analysis Report by Country (Thailand, India, Mexico, Costa Rica, Malaysia, Singapore, Brazil, Colombia, Turkey, Taiwan, South Korea, Spain, Czech Republic, China), and Segment Forecasts, 2022–2030*, n.d.

Health-Tourism.com. *The History of Medical Tourism*, 2020. Available at: www.health-tourism.com/ medical-tourism/history/ (accessed April 13, 2020).

Herrick, D. M. Medical Tourism: Global Competition in Health Care. *Natl. Center Policy Anal.* **2007,** *1*, 2–3.

Hong, Y. A. Vision 2.0 Medical Tourism and Telemedicine: A New Frontier of an Old Business. *J. Med. Internet Res.* **2016,** *18* (5), e115. doi:10.2196/jmir.5432. Available at: www. ncbi.nlm.nih.gov/pmc/articles/PMC4895094/ (accessed April 13, 2020).

Kannan, P. Dubai Reveals Master Plan for 500,000 Medical Tourists a Year. *The National*, 2014. http://www.thenational.ae/uae/tourism/dubai-reveals-master-plan-for-500-000-medical-tourists-a-year

Kucukusta, D.; Guillet, B. D. Measuring Spa-goers' Preferences: A Conjoint Analysis Approach. *Int. J. Hosp. Manage.* **2014,** *41*, 115–124.

Levebvre, B.; Bochatan, A. The Rebirth of the Hospital: Heterotopia and Medical Tourism in Asia. *Asia on Tour*; Routledge: London and New York, NY, 2002; pp 113–124.

Lobo, F. Industry 4.0—Manufacturing and the Future of Medical Things, 2020. Available at: www.asianhhm.com/technology-equipment/industry-manufacturing-future-medical-things (accessed April 13, 2020).

Lunt, N.; Hardey, M.; Mannion, R. Nip, Tuck and Click: Medical Tourism and the Emergence of Web-Based Health Information. *Open Med. Inform. J.* **2010,** *4*, 1–11.

Medical Technology. *Medical Tourism: How Is Digital Tech Reshaping the Industry?*, 2020. available at: https://medical-technology.nridigital.com/medical_technology_jan20/medical_tourism_how_is_digital_tech_reshaping_the_industry (accessed April 13, 2020).

Nehra, A.; Gettman, M.; Rivera, M.; Agarwal, D.; O'Neil, D.; Jenkins, S.; Viers, B. Patients Are Willing to Utilize Wearable Devices for Their Care: A Survey of Perceptions and Acceptance of Wearable Technology for Health Monitoring in a Urological Patient Population. *Urol. Pract.* **2017,** *4* (6), 508–514. doi:10.1016/j.urpr.2016.10.002.

Patrick, J.; Puterman, M. L. Reducing Wait Times through Operations Research: Optimizing the Use of Surge Capacity. *Healthcare Policy* **2008,** *3* (3), 75–88.

Qi, S.; Law, R.; Buhalis, D. Usability of Chinese Destination Management Organization Websites. *J. Travel Tour. Mark.* **2008,** *25* (2), 182–198.

Raman, R. K.; Varshney, L. R. Dynamic Distributed Storage for Scaling Blockchains. *ArXiv Preprint* **2017,** 1–19. ArXiv:1711.07617.

Sharp, G.; Tiggemann, M.; Mattiske, J. The Role of Media and Peer Influences in Australian Women's Attitudes towards Cosmetic Surgery. *Body Image* **2014,** *11* (4), 482–487. https://doi.org/10.1016/j.bodyim.2014.07.009

Singh, N. Exploring the Factors Influencing the Travel Motivations of US Medical Tourists. *Curr. Issues Tour.* **2013,** *16* (5), 436–454.
Sirdeshmuhk, D.; Singh, J.; Berry, S. Customer Trust, Value, and Loyalty in Relational Exchange. *J. Mark.* **2002,** *66*, 15–37.
Smith, A. K. Health Care Bargains abroad. *Kiplinger's Pers. Finan.* **2012,** *66* (1), 65–68.
Swain, D.; Sahu, S. Opportunities and Challenges of Health Tourism in India. In *Conference on Tourism in India—Challenges Ahead*, 15, 17, Citeseer, 2008.
Wong, K.M.; Musa, G. Medical Tourism in Asia: Thailand, Singapore, Malaysia, and India. In *Medical Tourism: The Ethics, Regulation, and Marketing of Health Mobility*; Hall, C. M., Ed.; Routledge: London and New York, NY, 2013; Chapter 11, pp 167–186.
Yusof, N.; Rosnan, H.; Zamzuri, N, Internationalisation Process of Medical Tourism Industry in Malaysia, a Sequential Approach. *Rev. Public.* **2019,** *6* (19), 448–462.

CHAPTER 16

RESOURCES AND EXPERIENCES TO AID IN MAKING THE RIGHT HEALTHCARE DECISIONS

ABSTRACT

Decades ago, the only source of information people had before opting to travel for treatment was a referral from their treating physician, and medical care was always a matter of blind faith and lack of transparency as patients were unable to check various healthcare parameters before making a medical journey. Now with the technological boom, people have access to resources like the internet where they can look for hospital quality, doctor expertise, and hospital infrastructure and be sure about their decision to travel. The availability of internet resources has completely changed the dynamic of decision-making in the last few years and made care better, transparent, and more satisfactory. This chapter explores the aids and resources that are helping in right healthcare decision making and also covers reasons why someone could go abroad for medical care.

A lot goes on while deciding and choosing the right physician for your health needs. It is widely accepted that patient participation in healthcare decisions is a sign of high-quality care (Entwistle et al., 2011). With the possibility to find and share firsthand and anecdotal knowledge about health and wellbeing because of the availability of the internet, the growth of social media, and peer resources, patients now have access to information on a range of health-related decisions (Yan et al., 2019; Witteman and Zikmund, 2012). This collaborative knowledge-building and information help them to get ready for doctor's appointments or to supplement them (Caiata-Zufferey et al., 2010).

Medical, Dental, and Wellness Tourism: A Postpandemic Perspective. Mary Schreiber Swenson, PhD, and Amit Bansal (Authors).

The greater patient involvement has allowed for improved decision-making by better equipping patients to work with healthcare practitioners to share information about their health and ask more questions (Townsend et al., 2015; Iverson et al., 2008). Healthcare personnel is interacting with patients more flexibly in tandem with this rising level of patient involvement. Together, these modifications represent a slow elimination of the conventional paternalistic healthcare model, in which healthcare consumers followed the advice of medical professionals blindly, to one that involves reciprocal engagement and shared health decisions (Met Internet, 2015). The opinions and situations of people and their families are essential considerations in this collaborative decision-making process.

Having the knowledge required to debate possibilities and preferences is a fundamental element of collaborative decision-making. Shared decision-making has traditionally been defined and researched in the setting of the consultation room (Rapley, 2008). In these situations, patients frequently participate in choosing their treatments and have sporadic dyadic interactions with their medical experts. However, decision-making processes are more complicated and varied, and instead of happening in isolated, discrete interactions, they are more likely to be changed over time by exposure to and reflection on a variety of resource contacts. People making decisions have access to many resources (both informative and supporting) at various times. This more dispersed idea can be used to frame Internet resources for health advice and information for informed decision-making.

With 72% of US-based health consumers and 75% of UK-based health consumers often looking online for information about illnesses, treatments, and medical procedures (Fox, 2019; Department of Health and UK Trade Investment, 2019), the internet has been recognized as a catalyst promoting the power of patients in terms of the availability of the right resources and ability to make educated choices (Department of Health, 2012). The influence and anonymity provided by the internet environment allow for the discussion of difficult subjects and give socially and geographically isolated people the chance to engage in health discussions (Reisner et al., 2018). When presented with a health issue, people are motivated to self-diagnose and research the condition (Best et al., 2016). They may do so by referring to conventional factual or statistical web sources as well as from other people's experiences (France et al., 2011; Ziebland and Herxheimer, 2008). Personal experience-based information is well-liked and widely used to convey social and emotional information about the methods and results of using the right resources that aid in determining the healthcare choices and help the prospecting patients with all the treatment options that are accessible (Entwistle et al., 2011).

A variety of web-based resources may help assist decision-making processes and influence final judgments, according to the research on the internet and health decision-making (Ziebland and Wyke, 2012; Lagan et al., 2010). Online support groups and peer-curated resources help people make decisions with more confidence and empowerment (Van Uden-Kraan et al., 2009). They also provide opportunities for people to find possibilities and reflect on previous decisions (Sillence and Bussey, 2017). The truth is that people will probably come into contact with both information that is based on facts and information that is based on patient experiences, (Blythe et al., 2017) and they will likely use both in various ways. Utilizing factual data, for instance, to support specific decisions and personal experiences to offer context and experiential detail to compare various options (Entwistle et al., 2011).

Decisions go beyond the doctor's consulting room and are changed throughout time by knowledge and interactions. Decisions are more likely to be ongoing processes that are periodically reviewed than discrete events. Depending on the type of health problem, this will change, resulting in different timeframes, additional resources, and supporting characters (involving family, friends, and other close individuals) (Montori, 2006).

Additionally, telemedicine and virtual second opinion services offered by certain key players like MediGence in the healthcare domain also allow patients the option to weigh their options, have a chat with an expert online, and then make a choice for their treatment. It is also possible to save the cost of travel by consulting a doctor virtually prior as they will guide the patient in the right direction by recommending the best course of treatment options which may or may not require them to travel. Review and rating platforms also play a key role in guiding patients in informed decision-making. Platforms like Trustpilot offer reviews of healthcare providers that can help patients get an overall picture of the services they can expect when they visit one. On the other hand, dedicated healthcare review and rating platforms like revU provides patients with a list of doctor/physicians from around the globe along with their reviews and ratings. The ratings are further broken down into various parameters making it deeply insightful for the prospecting patients to make an educated choice. Not just that, they can also look for hospitals and check hospital ratings where they wish to get treated.

Decades ago the only source of information people had before opting to travel for treatment was a referral from their treating physician. It was always a matter of blind faith and lack of transparency as patients were unable to check various healthcare parameters before making a medical journey. Now with the technological boom, people have access to resources like the internet where they can look for hospital quality, doctor expertise, and hospital infrastructure and be sure about their decision to travel. They can also compare costs with other countries providing similar quality of health and be really sure about their line of treatment and the treating physician. The technology has also provided patients with an option to search for doctors online and even book a teleconsultation with them prior to visiting them in person. This is crucial and one of the most impactful resources in making wise healthcare choices. Likewise, telehealth is giving a multitude of options to the patients to opt for primary care, second opinion, and rehabilitation care options. The availability of internet resources has completely changed the dynamic of decision-making in the last few years and made care better, transparent, and more satisfactory. Moreover, a humongous amount of information is now available on the internet which can provide a clear picture to patients about different countries, the quality of care provided by their healthcare facilities, political structure, healthcare standards, and details regarding the safety of the foreign nationals.

Alongside the era of technology, we cannot oversee the role of a good healthcare advisor. When a patient deals with a knowledgeable service provider, they can be of great assistance by helping them know all the details about the country where they wish to travel for treatment. Not just that, a patient advisor can also guide them with multiple options that have similar services and help in comparing costs to make the best decision. Furthermore, they take care of all the operative and administrative tasks giving patient's the space and time to focus on healing.

One commonly overlooked factor in healthcare is assurance. Many people choose the best hospital, doctor, and services while navigating through an enormous amount of resources but they overlook reconfirming if the proposed line of treatment is the right one. There are multiple resources in place to help a patient be assured and feel validated about their diagnosis. ThinkTwice by MediGence is one such platform where a global panel of skilled experts gets in touch with patient data, scans, history, and reports virtually to deliver the line of treatment options and confirm their diagnosis. After all, it is better to be sure than undergo a wrong treatment journey.

16.1 AIDS FOR DECISION-MAKING IN HEALTHCARE

When it comes to decision-making in healthcare various studies have come out with different results. Some claim that decisions where there is no clear optimal option and doctors are in a balance of interest, the best-suited way is to promote collaborative decision-making (Elwyn et al., 2000; Woolf, 1997). This type of decision-making is also termed "Preference-sensitive" (Wennberg et al., 2002; Whitney, 2003). Others have claimed that shared decision-making should be used more frequently because people frequently choose not to accept medical professionals' suggestions, in essence choosing to "do nothing" (Steven, 2001).

A decision aid's main function is to help people select a particular option from a range of alternatives to deal with a clinically relevant problem (O'Connor et al., 2003). Decision aids are designed to make it easier for patients and doctors to make decisions together (Barry, 2002; O'Connor et al., 1999). They are not meant to take the place of doctor–patient interaction, but rather to supplement it by offering, at the very least, comprehensive details regarding the various treatment alternatives and their expected results (Barry, 2002; O'Connor et al., 1999, 2003).

The layout and content of decision aids can vary greatly and can include details about the disease or condition, costs associated with various options, exercises to help users identify their preferences and values, and accounts of other people's decision-making experiences (often in the form of personal testimony), and guidance or instruction in decision-making or speaking with medical professionals (Barry, 2002; O'Connor et al., 2003). Nowadays, there are numerous consensus processes to establish standards by which a patient can assess the quality of a care facility, hospital, or doctor. Many healthcare facilities also aid patients by providing them with the physical forms of pamphlets or other guides or booklets available both online and offline to help in choosing a treatment plan for some specific procedures like knee replacement or ACL, care options, and check costs. However, the same is not the case everywhere around the world. People living in isolated or underdeveloped nations like Africa do not have access to many kinds of decision-making resources and aid. Nonetheless, with the advent of technology and the fast progressing adaptation of telehealth all around the world, patients now have the option to use digital resources to avoid any confusion and access the best quality of care within their desired geography.

The numerous ways that decision aids have been used include scripted personal counseling, handouts, booklets, audio-guided workbooks, interactive videodiscs, videotapes, and interactive websites (O'Connor et al., 2003).

Most are made to be viewed by patients before an appointment with a doctor. Now, it also includes virtual doctor consultations that can eliminate a lot of doubts before physically meeting one. Menorrhagia treatment, hepatitis B vaccination and screening, back surgery, atrial fibrillation treatment, dental surgery, infant vaccination schedules, menorrhagia screening, prostate and colon cancer treatment, prenatal screening, genetic testing, benign prostate disease treatment, hormone replacement therapy, ischemic heart disease treatment, and obstetrics are just a few of the topics that have been covered in detail in various forms of online and offline resources.

16.2 IMPORTANCE OF VARIOUS RESOURCES IN HEALTHCARE DECISION-MAKING

It was observed in certain studies that following exposure to some healthcare resources and aids, decisional conflict was generally reduced, with patients reporting feeling more educated as one of the biggest benefits (O'Connor et al., 2003). Numerous studies have discovered greater satisfaction with the clinical decision-making process and its results. People who saw a decision aid were less likely to stay unsure, and they frequently modified their preferences after reading the information. According to O'Connor et al. (2003), exposure also raised the chance that people would forgo major elective surgery in favor of more conservative alternatives.

Many people live in parts of the world where adequate treatment facilities are not available and many others live in places where although high-quality treatments and advanced medical technologies are available they are extremely costly. Many patients are unable to afford treatments at much higher costs and so they may check online resources and then choose the option to travel abroad for medical treatment. This problem is resolved with help of key healthcare players like MediGence who provide transparent costs by using cost calculators and help patients identify the most suitable option for them. Additionally, patients have the option to check doctor profiles, experience, and education. All of the healthcare data and records are managed by secured and safe portals ensuring the safety of medical information. All these factors together contribute to building confidence among patients regarding their healthcare choices. And after going through all the information and asking all the right questions using resources like reviews and ratings, online search guides, and patient testimonials, patients feel at ease when they choose to travel for treatment. For instance, millions of Americans travel to receive medical care each year. Mexico, Canada, as well as nations in Central America, and South America are popular destinations for medical tourists from the United States.

There are several reasons why someone could go abroad for medical care, including:

- Cost: The price of receiving care or undergoing a procedure that may be less expensive abroad.
- Culture: To obtain treatment from a medical professional who is fluent in the traveler's language and culture.
- Getting a procedure that is not available or approved in the residing country.
- Quality of care.

The most frequent treatments people receive when traveling for medical tourism include dental work, surgery, cosmetic surgery, fertility treatments, organ and tissue transplants, and cancer therapy. Online resources as well as hearing what other patients, who traveled before them to the same country, have to say can make a lot of difference in building trust and allowing patients to move forward with a treatment option without fear.

Patients are constantly seeking consultations and guidance. For instance, when you are in touch with the ideal healthcare provider, and you wish to travel and seek physical care and treatment under their guidance, at times you do not have a clear way to travel and you deal with many questions and steps that you may or may not know how to deal with on your own. With a professionally managed healthcare player like MediGence, you can get operational guidance on parameters ranging from health, level of technology, country-wise trends and technology, and every other detail that you may be seeking. All based on your illness and desired personal preferences

KEYWORDS

- **resources**
- **healthcare**
- **decision-making**
- **technology**
- **medical tourism**
- **travel for treatment**

REFERENCES

Barry, M. J. Health Decision Aids to Facilitate Shared Decision Making in Office Practice. *Ann. Intern. Med.* **2002,** *136*, 127–135.

Best, P.; Gil-Rodriguez, E.; Manktelow, R.; et al. Seeking Help from Everyone and No-One: Conceptualizing the Online Help-Seeking Process among Adolescent Males. *Qual. Health Res.* **2016,** *26*, 1067–1077.

Blythe, J. M.; Sillence, E.; Briggs, P. The Role of Trust in eHealth Websites: The Enduring Power of Impartiality. In *Frontiers in Public Health. Conference Abstract: 3rd UCL Centre for Behaviour Change Digital Health Conference*, University College London: London, February 22–23, 2017.

Caiata-Zufferey, M.; Abraham, A.; Sommerhalder, K.; et al. Online Health Information Seeking in the Context of the Medical Consultation in Switzerland. *Qual. Health Res.* **2010,** *20*, 1050–1061.

Department of Health. The Power of Information: Giving People Control of the Health and Care Information They Need. Department of Health: London, 2012. https://www.gov.uk/government/publications/giving-people-control-of-the-health-and-care-information-they-need (accessed July 8, 2019).

Department of Health and UK Trade and Investment. The UK: Your Partner for Digital Health Solutions, 2015. https://www.gov.uk/government/publications/the-uk-your-partner-for-digital-health-solutions (accessed April 1, 2019).

Elwyn, G.; Durand, M. A.; Song, J.; et al. A Three-Talk Model for Shared Decision Making: Multistage Consultation Process. *BMJ* **2017,** *6*, 359.

Elwyn, G.; Edwards, A.; Kinnersley, P.; Grol, R. Shared Decision Making and the Concept of Equipoise: The Competences of Involving Patients in Healthcare Choices. *Br. J. Gen. Pract.* **2000,** *50*, 892–899.

Entwistle, V. A.; France, E. F.; Wyke, S.; et al. How Information about Other People's Personal Experiences Can Help with Healthcare Decision-Making: A Qualitative Study. *Patient Educ. Couns.* **2011,** *85*, 291–298.

Fox, S. The Social Life of Health Information, 2014. http://www.pewresearch.org/fact-tank/2014/01/15/the-social-life-of-health-information/ (accessed July 8, 2019).

France, E. F.; Wyke, S.; Ziebland, S.; et al. How Personal Experiences Feature in Women's Accounts of Use of Information for Decisions about Antenatal Diagnostic Testing for Foetal Abnormality. *Soc. Sci. Med.* **2011,** *72*, 755–762.

Iverson, S. A.; Howard, K. B.; Penney, B. K. Impact of Internet Use on Health-Related Behaviors and the Patient-Physician Relationship: A Survey-Based Study and Review. *J. Am. Osteopath. Assoc.* **2008,** *108*, 699–711.

Lagan, B. M.; Sinclair, M.; George Kernohan, W. Internet Use in Pregnancy Informs Women's Decision Making: A Web-Based Survey. *Birth* **2010,** *37*, 106–115.

Montori, V. M.; Gafni, A.; Charles, C. A Shared Treatment Decision-Making Approach between Patients with Chronic Conditions and Their Clinicians: The Case of Diabetes. *Health Expect* **2006,** *9*, 25–36.

O'Connor, A. M.; Rostom, A.; Fiset, V.; Tetroe, J.; Entwistle, V.; et al. Decision Aids for Patients Facing Health Treatment or Screening Decisions: Systematic Review. *Br. Med. J.* **1999,** 319, 731–734.

O'Connor, A. M.; Stacey, D.; Entwistle, V.; Llewellyn-Thomas, H.; Rovner, D.; et al. Decision Aids for People Facing Health Treatment or Screening Decisions. *Cochrane Database Syst. Rev.* **2003,** CD001431.

Rapley, T. Distributed Decision Making: The Anatomy of Decisions-in-Action. *Sociol. Health Illn.* **2008,** *30*, 429–444.

Reisner, S. L.; Randazzo, R. K.; White Hughto, J. M.; et al. Sensitive Health Topics with Underserved Patient Populations: Methodological Considerations for Online Focus Group Discussions. *Qual. Health Res.* **2018,** *28*, 1658–1673.

Sillence, E.; Bussey, L. Changing Hospitals, Choosing Chemotherapy and Deciding You've Made the Right Choice: Understanding the Role of Online Support Groups in Different Health Decision-Making Activities. *Patient Educ. Couns.* **2017,** *100*, 994–999.

Steven, K. Shared Decision-Making. *Br. J. Gen. Pract.* **2001,** *51*, 61–62.

Townsend, A.; Leese, J.; Adam, P.; et al. eHealth, Participatory Medicine, and Ethical Care: A Focus Group Study of Patients' and Health Care Providers' Use of Health-Related Internet Information. *J. Med. Internet Res.* **2015,** *17*, e155.

Van Uden-Kraan, C. F.; Drossaert, C. H.; Taal, E.; et al. Participation in Online Patient Support Groups Endorses Patients' Empowerment. *Patient Educ. Couns.* **2009,** *74*, 61–69.

Wennberg, J. E.; Fisher, E. S.; Skinner, J. S. Geography and the Debate over Medicare reform. *Health Aff. (Millwood) Suppl. Web Exclus.* **2002,** W96–W114.

Whitney, S. N. A New Model of Medical Decisions: Exploring the Limits of Shared Decision Making. *Med. Decis. Mak.* **2003,** *23*, 275–280.

Witteman, H. O.; Zikmund-Fisher, B. J. The Defining Characteristics of Web 2.0 and their Potential Influence in the Online Vaccination Debate. *Vaccine* **2012,** *30*, 3734–3740.

Woolf, S. H. Shared Decision-Making: The Case for Letting Patients Decide Which Choice Is Best. *J. Fam. Pract.* **1997,** *45*, 205–208.

Yan, L.; Yan, X.; Tan, Y.; et al. Shared Minds: How Patients Use Collaborative Information Sharing Via Social Media Platforms. *Prod. Oper. Manage.* **2019,** *28*, 9–26.

Ziebland, S.; Herxheimer, A. How Patients' Experiences Contribute to Decision Making: Illustrations from DIPEx (Personal Experiences of Health and Illness). *J. Nurs. Manage.* **2008,** *16*, 433–439.

Ziebland, S.; Wyke, S. Health and Illness in a Connected World: How Might Sharing Experiences on the Internet Affect People's Health? *Milbank Q.* **2012,** *90*, 219–249.

CHAPTER 17

TELEMEDICINE REVOLUTIONIZING THE MEDICAL INDUSTRY

ABSTRACT

The ability of healthcare professionals to treat many patients while not being physically present has improved because of telemedicine. Today, telemedicine growth is increasing, more rapidly than any other vertical of the medical industry, especially as the demand is higher from mobile users. Patients browse through internet healthcare sites that are delivering teleconsultations, and it helps the new patients in locating the specialist doctor that is ideally suited to their needs, and they also have the option to check their credentials before booking a consultation promoting transparency. In many developed nations, there are nursing care centers that offer advice via virtual modes to patients who are self-treating at home. Although the initial generation of telemedicine technologies introduced many providers to telehealth, the new wave will have much more to offer. This chapter talks about digital health monitoring, remote delivery of healthcare for elderly and disabled, and telerehabilitation.

Coronavirus disease in 2019 has posed a threat to the way we currently conduct, provide, and receive modern medicine (Klompas, 2020; Bhaskar et al., 2020), raising a need to shift to other, better alternatives for delivering care. Telemedicine emerged as one of the most stable delivery methods in healthcare and even postpandemic it is expected to keep blooming. The American Telemedicine Association defines telemedicine as "the remote delivery of health information and healthcare services through telecommunications technology" (American Telemedicine Association, 2020). It enables patient care while limiting the need for physical contact, hence lowering

Medical, Dental, and Wellness Tourism: A Postpandemic Perspective. Mary Schreiber Swenson, PhD, and Amit Bansal (Authors).

the load on healthcare facilities and the spread of infections. It can be used for medication compliance, ongoing management of chronic illnesses, doctor–patient consultation, and other virtual services (Bhaskar et al., 2020). Through telehealth platforms and supporting technologies like the Internet of Things (IoT), larger populations can be served at once eliminating the need to travel for treatment and saving a large amount of healthcare spending.

Digital and telemedicine technologies have the ability to significantly improve access and care in rural areas. Many medical subspecialties have switched from in-person outpatient care to remote tele- or e-health due to the requirement for health systems to quickly adapt to new needs. In general, remote patient assessment is a great alternative instead of having patients visit primary care or outpatient clinics. During a pandemic, this proved to be very useful in ensuring patient continuity for people with chronic conditions (Bhaskar et al., 2020). Additionally, many patients are weak and therefore are at significant risk for infection and adverse consequences, such as mortality. Notably, telemedicine also reduces the risk of infection exposure for medical professionals, permits talks among multidisciplinary teams, and allows quick access to subspecialists who are not immediately present in person.

Patients that exhibit infectious symptoms can be assessed, and patients who are housebound, such as the elderly or weak, may be candidates for telemonitoring via video consultations. Patients who are immunocompromised, single parents, or those who depend on others for transportation are just a few examples of people with limited access to healthcare who cannot resort to treatment methods that require traveling to the clinic and may benefit from telemedicine. Through the use of telemedicine, patients could be monitored while receiving home rehabilitation programs, physiotherapy, psychological counseling, social work consultations, and speech and language therapies remotely.

In general, telehealth has the potential to improve the efficiency, coordination, and accessibility of healthcare. This area is still in its initial stages and expanding with great speed. For instance, the use of video consultations and telemonitoring patients' vital signs and symptoms has effectively reduced the cases of mortality, reduced the chances of hospitalization, and as a result increased patients' quality of life. People opt for diagnosis and recovery plans for several good reasons. They make them feel confident about receiving the right care and the best quality treatment. Telemedicine has proven to be a great source for imparting mental health treatments as it eliminates some of the reasons that stop patients from going to a psychiatrist or receiving this critical form of treatment (Wilson and Maeder, 2015; Hajesmaeel-Gohari and Bahaadinbeigy, 2021; Lupton and Maslen, 2017).

Patients can obtain medical care via telemedicine at the mutual convenience of their doctor. This could imply that patients no longer need to worry about scheduling daycare or taking time off from work for visiting a doctor. The ability of healthcare professionals to treat many patients while not physically present has improved because of telemedicine. Patients are now more willing to consider video consultations as a superior option. Although the initial generation of telemedicine technologies introduced many providers to telehealth, the new wave will have much more to offer. For instance, clinicians can utilize natural language processing to automatically take notes. Specialists will be able to provide their valuable input from a distance during emergency operations. A healthcare provider may upload the data gathered by medical equipment to an IoT cloud platform, where it is consolidated. The IoT systems used by healthcare professionals for patient management will then receive this information. Artificial Intelligence is a recent development in telemedicine technologies that can assist doctors in performing their jobs more efficiently. With wearables and other remote patient tracking tools, technology will stay in the foreground and keep patients informed (Chunara et al., 2021; Flumignan et al., 2019; Kaspar, 2013).

For patients in rural places, telemedicine technology has immense potential. The biggest impact is seen in a number of nations where there are little or no healthcare facilities. For instance, the African population is deprived of the right to healthcare due to the lack of physicians available for providing the treatment. With the help of telemedicine, it is possible for the patients to connect with a physician outside of their geographical limitations, eliminating this issue to a large extent. All patients and doctors must have adequate hardware and software security in place to guarantee an accurate medical history. Through internet video conferencing, several clinics offer doctor virtual appointments. These appointments allow individuals to see a doctor without a need for an in-person visit.

As part of their healthcare offerings, many major key players like MediGence are offering telehealth services in various crucial treatment areas like oncology, cosmetic care, cardiology, dermatology, orthopedics, and so on. The patient can directly access doctors from global destinations and get consultations for their conditions. It is not only possible to take primary consultations but the patients now have the option to get a second opinion virtually as well where the specialist will explain the diagnosis and line of treatment to the patient in detail. Along with a second opinion, online rehabilitation, and recovery care options are also available for the patients.

17.1 DIGITAL HEALTH MONITORING

Healthcare surveillance is essential to preserve healthcare quality. Smart connectivity systems have been developed with the incorporation of digital health-tracking technology and services. With the help of direct video conferencing, this technology paves the way for patient control and a far deeper understanding of patients. Doctors can maintain a catalog of patient records that helps in saving valuable time. It increases the treatment process's accuracy between the doctor and patient. Cloud-based medical record-keeping of telemedicine data with patient information and recorded video consultation is frequently used to assure data safety. There are many secure channels to ensure data privacy digitally and help in effective data monitoring (Dalley and Rahman, 2021; Martínez et al., 2004; Kyriacou et al., 2003).

In many developed nations, there are nursing care centers that offer advice via virtual modes to patients who are self-treating at home (Rockwell and Gilroy, 2020; Bashshur et al., 2011; Funderskov et al., 2019). This is another effective way in which telemedicine is lucratively stepping up in the medical industry. This method entails a question–answer format to give your care provider an idea about your condition. People can use this technology to remember their appointments, refill prescriptions, take blood pressure medication, and so on. Additionally, patients can participate in step-by-step training services catered to their particular ailment, complete a series of self-tests, and e-mail their symptoms to doctors. By simply putting care monitoring applications and cell phones in patients' hands, electronic health technology simplifies the management of chronic illnesses in all situations (Lokkerbol et al., 2014; Charles, 2000).

Medical records can be easily shared with any professional with expertise in treating a particular patient's ailment, even a patient's caretaker who requires the information for insurance requirements can easily obtain the information digitally via secure channels. Patients can communicate with their doctors through teleconference to ask inquiries regarding their symptoms. In-person assessments are being replaced by photos and video consultations to enable real-time cooperation and remote patient diagnostics. Tracking medical progress and performing inspections as needed, saves time for both patients and doctors. The telehealth platforms guarantee easy patient record gathering and consolidation in one place. The quicker delivery of healthcare services benefits customers, treating doctors, and pharmacies. Doctors can write and send prescriptions more accurately with the help of written digital reports. A lot of time is saved, connectivity is enhanced, and customer happiness increases. It's a win–win for all (Yellowlees et al., 2010; Wernhart et al., 2019; Javaid et al., 2021; Ahmad et al., 2021).

Today, telemedicine growth is increasing, more rapidly than any other vertical of the medical industry, especially as the demand is higher from mobile users. Patients browse through internet healthcare sites that are delivering teleconsultations, and it helps the new patients in locating the specialist doctor that is ideally suited to their needs. They also have the option to check their credentials before booking a consultation promoting transparency (Pooni et al., 2021; Omboni et al., 2020).

Applications in telemedicine offer the potential to expedite the treatment of many different medical ailments. When a patient receives advice, opinion, and medication from a qualified physician who provides thorough knowledge about their symptoms, it builds their confidence. Telemedicine is a way to keep monitoring a patient's condition continuously. Unequivocally, using cutting-edge services and technologies has also increased access to healthcare facilities. Any hospital or treatment facility can immediately get in touch and access medical specialists, and consultants, and tap into their education to improve their deliverables. The hospitals and clinics from one country can easily collaborate or get in association with the services of doctors and other healthcare providers from another geographical location in a short amount of time. This opens up multiple opportunities and channels to get treatment for a patient eliminating the lingering problem of accessibility in healthcare (Mihova et al., 2009; Eisenstein et al., 2020).

An important example of this is the UK healthcare system. Considered to be one of the best healthcare delivery systems, the National Health Service (NHS) usually lacks in providing timely treatment to the patient as there are always long wait ups and months' worth of appointments pending and limited availability of doctors in the country. With help of telehealthcare systems and web-based monitoring, the hospitals can get additional support and reduce the burden on the current system. It is predicted that new hospital support and web-based, digital general practice consultations will cut down on outpatient sessions by up to a third (NHS, 2019). This might decrease the number of visits to hospitals by 30 million patients yearly, saving the NHS more than \$1.35 billion approx. in new expenses every year (NHS, 2019). Although the management of chronic diseases (such as diabetes) and perioperative care (such as orthopedic surgery) have traditionally been the focus of remote patient monitoring (RPM), the expansion of RPM to wider patient groups promises additional benefits to both patients and healthcare professionals. RPM's clinical outcomes are not well supported by evidence, but its economic advantages are clearly demonstrated; for instance, there will be a decrease in the workload of physicians with the redirection of services to environments that are highly demanding.

Around the world, cardiology is one of the specialties where comprehensive telemedicine systems have been applied successfully. ECG with Holter monitoring, echocardiography records, and virtual auscultation have all been made possible by the monitoring of heart rhythm in patients using implanted or real-time wearable equipment. Cardiologists are worried about the new information that suggests patients with COVID-19 have heart problems (Bhaskar et al., 2020; American Telemedicine Association, 2020; World Health Organization, 2018; Sharma et al., 2017; Fried et al., 2020).

This covers heart-related issues such as arrhythmias (Shi et al., 2020; Inciardi et al., 2020), cardiac damage, heart failure, myocarditis, pericarditis, and vasculitis. Additionally, patients with preexisting cardiovascular diseases who get COVID-19 have extremely bad results, and a 5- to 10-fold increase in death rate (Bonow et al., 2020). Therefore, the American government has promoted remote monitoring and virtual visits for patients with heart issues since COVID-19 (Bhatt et al., 2020).

In a statement, the American Heart Association (AHA) highlighted the use of telemedicine in pediatric cardiology through the use of cutting-edge video technologies like fetal echocardiography for prenatal diagnosis, confirmatory echo tests, external rhythm monitoring, and catheterization lab, and personal tele-electrophysiology (Satou et al., 2017). Notably, noninvasive telemonitoring of patients with heart failure lower hospitalizations and all-cause mortality while also enhancing patient quality of life (Inglis et al., 2015).

17.2 REMOTE DELIVERY OF CARE FOR ELDERLY AND DISABLED

Telecare is being utilized more frequently to help elderly people preserve their independence and continue living in their own homes as the population ages and the prevalence of chronic health conditions rises (Barlow et al., 2007). Elderly people have emerged as one of the primary target demographics for telecare technology, with numerous devices readily available for those with ongoing medical disorders as well as for those with mobility issues or memory issues typical of advanced age. Population trends show that the world's population is aging quickly, notably in Europe. More people than ever before are getting older in many countries, especially in the more developed parts of the world. The percentage of people over the age of 60 is expected to rise from roughly 11% to 22% between 2000 and 2050.

The number of people over 80 is increasing quickly, and it is predicted that between 2000 and 2050, this number will quadruple (World Health Organization, 2014). The percentage of elderly persons, those over 65, in Europe is predicted to rise from 18.2% in 2013 to 28.1% in 2050. In Italy and Germany, 20.8% and 20.6% of the population were 65 or older in 2012 and have the oldest populations among the individual nations. Japan and many European nations have numbers that are almost as high (European Commission, 2014). In contrast, the population over the age of 65 in the USA makes up 14.5% of the total population. However, by 2030, this percentage will have increased to around 20% (United States Demographics Profile, 2014).

There is considerable doubt that shifts in age distribution have significant effects on people's individual and societal health, social conditions, and economic well-being. With an aging population comes a large rise in chronic disease prevalence and its consequences, necessitating greater care and welfare. Older persons are more prone to have several comorbidities and take multiple medications. The percentage of seniors who currently have at least two chronic illnesses coexisting ranges from 55% to 98% and the frequency of multimorbidity among the aged is predicted to rise. As a result, seniors will have irreversible functional loss, disability, and low quality of life (Marengoni et al., 2011; Rechel et al., 2013). Dementia, stroke, vascular complications of atherosclerosis, osteoarthritis, cancers, incontinence, osteoporosis, falls, fractures, and bedsores are age-dependent diseases that frequently occur during the aging process (Rosenthal et al., 2006). These illnesses cause physical and mental degeneration, necessitating constant supervision and frequent contact with medical and social care facilities.

For many seniors, using standard healthcare services—where the patient goes to a medical institution alone—becomes difficult, if not impossible. This is especially true if they live alone and are dependent exclusively on themselves. According to statistics, the annual cost to the healthcare system for diagnosing and treating people with chronic diseases has climbed dramatically in the majority of highly industrialized nations. In most healthcare systems, 5% of patients with various chronic illnesses account for 50% of the costs (Berk and Monheit, 2001). Additionally, it is expected that the expense of caring for senior people with chronic illnesses would increase significantly over the next few decades, with Alzheimer's care alone potentially costing more than $1 trillion a year (US Census Bureau, 2014).

Care provided through telemedicine is meant to address the unique requirements and expectations of senior citizens. The most popular types of telecare for older people are those based in the home. These systems typically offer three main services:

- safety and security monitoring, such as using sensors and detectors to keep a check on their health;
- monitoring health parameters and vital signs, like heart rate, blood pressure, body temperature, and glucose levels; and
- support via information and communication technology (Internet, telephone), like using teleconsultations and short messaging services (Barlow et al., 2007).

Monitoring patients with chronic illnesses, such as heart failure, hypertension, diabetes, asthma, chronic obstructive pulmonary disease, or stroke, is achievable with help of telecare technologies (PACITA Project, 2014; Parè et al., 2007; Clark and Gropen, 2015). The purpose of telemonitoring is to spot disorders, functional decline, and other significant changes in a patient's health before they need immediate attention in an ER or hospital, or long-term care in a nursing home. Chronic disease home telemonitoring appears to be a promising method of patient care that generates accurate and trustworthy data, empowers patients, affects their attitudes and actions, and may even improve their medical conditions (Takahashi et al., 2010).

In a systematic review conducted by Parè et al., patients were seen to be following telemonitoring programs and technology regardless of their nationality, socioeconomic status, or age. Noel et al. (2004) found that patients who used a home-based telemonitoring system that could measure vital signs, blood glucose levels, auscultation of the heart and lungs, and three-lead electrocardiography, pulse oximetry, and pain assessment had fewer bed days and emergency department visits.

Proactive healthy behaviors are formed through telemedicine-based care. In an emergency case, telemonitoring can identify pathological symptoms and abnormal test findings earlier than during/before a routine medical visit or inquiry, enabling the early implementation of the necessary preventive actions (Tomita et al., 2010). Telemonitoring has a significant impact on education. Patients who are conscious of their health status frequently begin to educate themselves on their disease and how to self-manage it. They have a better understanding of their body and are more motivated to modify unhealthy behaviors and live healthier lifestyles (Bashshur and Shannon, 2009).

Telecare lessens socioeconomic and regional inequities in access to care as well as the uneven distribution of care's quality. The ability of telemedicine systems to be electronically connected across geographical boundaries is one of the primary features. With the help of telecare technologies, older individuals can easily connect with a variety of functionally and geographically dispersed healthcare professionals at times that are convenient for both the patient and the healthcare provider. Doctors in varied practice settings can easily access evidence-based medicine and effective clinical decision-making tools, such as knowledgeable colleagues in tertiary care facilities.

Additionally, patients can get the right kind of treatment close to where they live, which is crucial for older individuals with chronic illnesses or disabilities because it affects their quality of life and level of care (Hanson et al., 2007). Last but not least, it should be remembered that telecare can assist patients and their families/caregivers who have significant mental and/or physical disabilities (Solli et al., 2015). A stroke victim or a patient with increasing dementia who is confused, scared, or suspicious and unable to carry out even the most basic daily duties. In such cases, teleconsultation and online monitoring are a boon that can help the patients in dealing with their health issues effectively without the need for relocation or travel.

17.3 TELEMEDICINE-ASSISTED REHABILITATION

Telemedicine can help with rehabilitation following acute events like stroke and traumatic brain injury (Bhaskar et al., 2020) as well as chronic illnesses like COPD, CVD, diabetes, and obesity that require continual rehabilitation efforts (Bhaskar et al., 2020). Studies have shown that stroke telerehabilitation systems with consultations, exercises, games, and therapy components have produced beneficial effects like enhancing patients' mental and functional well-being. Increased patient motivation and comfort as a result of being in a home environment are additional advantages (Chen et al., 2020).

Practitioners including physiotherapists, occupational therapists, speech and language therapists, and podiatrists typically carry out hands-on thorough movement assessments. Many patients were unable to receive any face-to-face rehabilitation during the COVID-19 epidemic due to the disruption of medical services and the protection of the most vulnerable (World Health

Organization, 2020; Gutenbrunner et al., 2020). As a result, professionals changed their methods of practice to include telerehabilitation—the provision of rehabilitation services through information and communication technology (Brennan et al., 2010). In the UK and across the globe, the pandemic led to a significant rise in the use of telephone and video-based consultations for rehabilitation assessments and interventions (Wade, 2021; Peek et al., 2020). Although efficacy has not yet been shown, systematic review data indicates that patient outcomes may be improved by these services just as well as face-to-face interventions (Cottrell et al., 2017; Tchero et al., 2018). When compared to standard face-to-face care, one review found that physiotherapy administered via video or telephone for a variety of musculoskeletal problems was linked with similar or better improvements in physical function and pain outcomes.

Another evaluation found that stroke patients' health-related quality of life had improved similarly in the control and telerehabilitation groups (Cottrell et al., 2018). Many patients with physical disabilities—including those with severe expressive communication difficulties (Styles, 2008), stroke (Galloway et al., 2014), and chronic musculoskeletal illnesses—perceive telerehabilitation to be acceptable. According to patient surveys, remote delivery of services may be favored because of its ease and reduced travel time (Bradwell et al., 2021; Tenforde et al., 2020).

Remote delivery of rehabilitation services also has the potential to minimize costs for health and social care providers, including the time and travel expenses of practitioners and their patients as well as the use of outpatient resources (Bini and Mahajan, 2017).

In many rehabilitation sectors, virtual reality is emerging as a novel technique (Holden, 2005). It has so far been used for vestibular therapy, upper limb rehabilitation (Henderson et al., 2007), physical, and learning disabilities (Yalon-Chamovitz and Weiss, 2007), and cognitive impairment. Telerehabilitation may have significant advantages, especially in terms of better communication and remote access to medical care. The sharing of knowledge and medical data is made possible by improved communication, which benefits patients, families, carers, physicians, and researchers. People who live in distant places now have the opportunity to obtain healthcare services assisting families caring for patients with low responsiveness.

For those with severe disabilities such as severe traumatic brain injury, chronic condition following stroke, advanced multiple sclerosis, and so on, the ability to communicate with rehabilitation facilities via telesystems is important. In these situations, remote communication with the physician, the nurses, and the therapist may be crucial in addressing and resolving the issues and reducing the need for hospitalization. In order to maintain the restored function after the completion of the rigorous rehabilitation programs and monitor the activity of the injured/operated organ, contact established with a professional via a telechannel means everything. Some patients who travel cross-borders to get treatment are often concerned about getting the right posttreatment rehabilitation in their home country. With help of telerehabilitation, this problem has been successfully eliminated. Patients can go back to their hometown and still consult their treating rehabilitation specialist via videoconferencing channels. Some key healthcare players like MediGence make it possible. The specialist stays with you throughout the session. You will require a helper throughout the process who will guide you while listening to the medical expert and watching their actions on a digital medium. Additionally, the platform strictly adheres to TEMOS and HIPAA compliances ensuring the safety of patient data.

KEYWORDS

- **telemedicine**
- **medical industry**
- **healthcare**
- **remote care**
- **virtual healthcare**
- **telehealth**

REFERENCES

Ahmad, R. W.; Salah, K.; Jayaraman, R.; Yaqoob, I.; Ellahham, S.; Omar, M. The Role of Blockchain Technology in Telehealth and Telemedicine. *Int. J. Med. Inf.* **2021,** *148*, 104399.

American Telemedicine Association. *About Telemedicine*, n.d. Available online at: http://legacy.americantelemed.org/main/about/about-telemedicine/telemedicine-faqs (accessed April 16, 2020).

Barlow, J.; Singh, D.; Bayer, S.; Curry, R. A Systematic Review of the Benefits of Home Telecare for Frail Elderly People and Those with Long-Term Conditions. *J. Telemed. Telecare* **2007,** *13*, 172–179.

Bashshur, R. L.; Shannon, G. W. National Telemedicine Initiatives: Essential to Healthcare Reform. *Telemed. J. E. Health* **2009,** *15*, 600–610.

Bashshur, R.; Shannon, G.; Krupinski, E.; Grigsby, J. The Taxonomy of Telemedicine. *Telemed. e-Health* **2011,** *17* (6), 484–494.

Berk, M. L.; Monheit, A. C. The Concentration of Health Care Expenditures, Revisited. *Health Aff. (Millwood)* **2001,** *20*, 9–18.

Bhaskar, S.; Bradley, S.; Israeli-Korn, S.; Menon, B.; Chattu, V. K.; Thomas, P.; et al. Chronic Neurology in COVID-19 Era: Clinical Considerations and Recommendations from the REPROGRAM Consortium. *Front. Neurol.* **2020a,** *11*, 664. doi:10.3389/fneur.2020.00664.

Bhaskar, S.; Rastogi, A.; Chattu, V. K.; Adisesh, A.; Thomas, P.; Alvarado, N.; et al. Key Strategies for Clinical Management and Improvement of Healthcare Services for Cardiovascular Disease and Diabetes Patients in the Coronavirus (COVID-19) Settings: Recommendations from the REPROGRAM Consortium. *Front. Cardiovasc. Med.* **2020b,** *7*, 112. doi:10.3389/fcvm.2020.00112.

Bhaskar, S.; Sharma, D.; Walker, A. H.; McDonald, M.; Huasen, B.; Haridas, A.; et al. Acute Neurological Care in the COVID-19 Era: The Pandemic Health System REsilience PROGRAM (REPROGRAM) Consortium Pathway. *Front. Neurol.* **2020c,** *11*, 579. doi:10.3389/fneur.2020.00579.

Bhatt, A.; Freeman, A.; Mullen, B. Telehealth: Rapid Implementation for Your Cardiology Clinic. Available online at: https://www.acc.org/latest-in-cardiology/articles/2020/03/01/08/42/feature-telehealth-rapid-implementation-for-your-cardiology-clinic-coronavirus-disease-2019-COVID-19 (accessed April 16, 2020).

Bini, S.; Mahajan, J. Clinical Outcomes of Remote Asynchronous Telerehabilitation Are Equivalent to Traditional Therapy Following Total Knee Arthroplasty: A Randomized Control Study. *J. Telemed. Telecare* **2017,** *23* (2), 239–247.

Bonow, R. O.; Fonarow, G. C.; O'Gara, P. T.; Yancy, C. W. Association of Coronavirus Disease 2019 (COVID-19) with Myocardial Injury and Mortality. *JAMA Cardiol.* **2020**. doi:10.1001/jamacardio.2020.1105.

Bradwell, H. L.; Baines, R. L.; Edwards, K. J.; Stevens, S. G.; Atkinson, K.; Wilkinson, E.; et al. Exploring Patient Experiences of Video Consultations During COVID-19 in an Outpatient Care Setting Using Routine Feedback Data from 955 Contacts. *medRxiv*. Preprint posted online on May 14, 2021.

Brennan, D.; Tindall, L.; Theodoros, D.; Brown, J.; Campbell, M.; Christiana, D.; et al. A Blueprint for Telerehabilitation Guidelines. *Int. J. Telerehabil.* **2010,** *2* (2), 31–34.

Charles, B. L. Telemedicine Can Lower Costs and Improve Access. *Healthc. Financ. Manage.* **2000,** *54* (4), 66.

Chen, Y.; Chen, Y.; Zheng, K.; Dodakian, L.; See, J.; Zhou, R.; et al. A Qualitative Study on User Acceptance of a Home-Based Stroke Telerehabilitation System. *Top Stroke Rehabil.* **2020,** *27*, 81–92. doi:10.1080/10749357.2019.1683792.

Chunara, R.; Zhao, Y.; Chen, J.; Lawrence, K.; Testa, P. A.; Nov, O.; Mann, D. M. Telemedicine and Healthcare Disparities: A Cohort Study in a Large Healthcare System in New York City during COVID-19. *J. Am. Med. Inf. Assoc.* **2021,** *28* (1), 33–41.

Clark, M. L.; Gropen, T. Advances in the Stroke System of Care. *Curr. Treat. Options Cardiovasc. Med.* **2015,** *17*, 355.

Cottrell, M. A.; Galea, O. A.; O'Leary, S. P.; Hill, A. J.; Russell, T. G. Real-Time Telerehabilitation for the Treatment of Musculoskeletal Conditions Is Effective and Comparable to Standard Practice: A Systematic Review and Meta-analysis. *Clin. Rehabil.* **2017,** *31* (5), 625–638.

Cottrell, M. A.; Hill, A. J.; O'Leary, S. P.; Raymer, M. E.; Russell, T. G. Patients Are Willing to Use Telehealth for the Multidisciplinary Management of Chronic Musculoskeletal Conditions: A Cross-sectional Survey. *J. Telemed. Telecare* **2018,** *24* (7), 445–452.

Dalley, D.; Rahman, R.; Ivaldi, A. Health Care Professionals' and Patients' Management of the Interactional Practices in Telemedicine Videoconferencing: A Conversation Analytic and Discursive Systematic Review. *Qual. Health Res.* **2021,** *31* (4), 804–814.

Eisenstein, E.; Kopacek, C.; Cavalcante, S. S.; Neves, A.C.; Fraga, G. P.; Messina, L. A. Telemedicine: A Bridge over Knowledge Gaps in Healthcare. *Curr. Pediatr. Rep.* **2020,** *8*, 93–98.

European Commission. *Population Structure and Ageing*; Eurostat, the Statistical Office of the European Union, May 2014. Available from: http://ec.europa.eu/eurostat/statistics-explained/index.php/Population_structure_and_ageing (accessed December 15, 2014).

Flumignan, C. D.; Rocha, A. P.; Pinto, A. C.; Milby, K. M.; Batista, M. R.; Atallah, Á. N.; Saconato, H. What do Cochrane Systematic Reviews Say about Telemedicine for Healthcare? *Sao Paulo Med. J.* **2019,** *137* (2), 184–192.

Fried, J. A.; Ramasubbu, K.; Bhatt, R.; Topkara, V. K.; Clerkin, K. J.; Horn, E.; et al. The Variety of Cardiovascular Presentations of COVID-19. *Circulation* **2020,** *141*, 1930–1936. doi:10.1161/CIRCULATIONAHA.120.047164.

Funderskov, K. F.; Boe Danbjørg, D.; Jess, M.; Munk, L.; Olsen Zwisler, A. D.; Dieperink, K. B. Telemedicine in Specialised Palliative Care: Healthcare Professionals and Their Perspectives on Video Consultations—A Qualitative Study. *J. Clin. Nurs.* **2019,** *28*, 3966–3976.

Galloway, M.; Marsden, D. L.; Callister, R.; Nilsson, M.; Erickson, K. I.; English, C. The Feasibility of a Telehealth Exercise Program Aimed at Increasing Cardiorespiratory Fitness for People after Stroke. *Int. J. Telerehabil.* **2019,** *12* (2), 9–28.

Gutenbrunner, C.; Stokes, E.; Dreinhöfer, K.; Monsbakken, J.; Clarke, S.; Côté, P.; et al. Why Rehabilitation Must Have Priority during and after the COVID-19-Pandemic: A Position Statement of the Global Rehabilitation Alliance. *J. Rehabil. Med.* **2020,** *52* (7), jrm00081.

Hajesmaeel-Gohari, S.; Bahaadinbeigy, K. The Most Used Questionnaires for Evaluating Telemedicine Services. *BMC Med. Inf. Decis. Making* **2021,** *21* (1), 1–1.

Hanson, J.; Percival, J.; Aldred, H.; Brownsell, S.; Hawley, M. Attitudes to Telecare among Older People, Professional Care Workers and Informal Carers: A Preventative Strategy or Crisis Management? *Univers. Access Inf. Soc.* **2007,** *6*, 193–205.

Henderson, A.; Korner-Bitensky, N.; Levin, M. Virtual Reality in Stroke Rehabilitation: A Systematic Review of Its Effectiveness for Upper Limb Motor Recovery. *Top Stroke Rehabil.* **2007,** *14*, 52–61.

Holden, M. K. Virtual Environments for Motor Rehabilitation: Review. *Cyberpsychol. Behav.* **2005,** *8*, 187–211; discussion 212–219.

Inciardi, R. M.; Lupi, L.; Zaccone, G.; Italia, L.; Raffo, M.; Tomasoni, D.; et al. Cardiac Involvement in a Patient with Coronavirus Disease 2019 (COVID-19). *JAMA Cardiol.* **2020,** *5*, 819–824. doi:10.1001/jamacardio.2020.1096.

Inglis, S. C.; Clark, R. A.; Dierckx, R.; Prieto-Merino, D.; Cleland, J. G. Structured Telephone Support or Non-invasive Telemonitoring for Patients with Heart Failure. *Cochrane Database Syst. Rev.* **2015,** Cd007228. doi:10.1002/14651858.CD007228.pub3.

Javaid, M.; Haleem, A.; Singh, R. P.; Suman, R. Substantial Capabilities of Robotics in Enhancing Industry 4.0 Implementation. *Cognit. Robot.* **2021,** *1*, 58–75.

Kaspar, B. J. Legislating for a New Age in Medicine: Defining the Telemedicine Standard of Care to Improve Healthcare in Iowa. *Iowa Law Rev.* **2013,** *99*, 839.

Klompas, M. Coronavirus Disease 2019 (COVID-19): Protecting Hospitals from the Invisible. *Ann. Int. Med.* **2020,** *172*, 619–620. doi:10.7326/M20-0751.

Kyriacou, E.; Pavlopoulos, S.; Berler, A.; Neophytou, M.; Bourka, A.; Georgoulas, A.; Anagnostaki, A.; Karayiannis, D.; Schizas, C.; Pattichis, C.; Andreou, A. Multi-purpose HealthCare Telemedicine Systems with Mobile Communication Link Support. *Biomed. Eng. Online* **2003,** *2* (1), 1–2.

Lokkerbol, J.; Adema, D.; Cuijpers, P.; Reynolds, C.F., III; Schulz, R.; Weehuizen, R.; Smit, F. Improving the Cost-effectiveness of a Healthcare System for Depressive Disorders by Implementing Telemedicine: A Health Economic Modeling Study. *Am. J. Geriatr. Psychiatr.* **2014,** *22* (3), 253–262.

Lupton, D.; Maslen, S. Telemedicine and the Senses: A Review. *Sociol. Health Illness* **2017,** *39* (8), 1557–1571.

Marengoni, A.; Angleman, S.; Melis, R.; et al. Aging with Multimorbidity: A Systematic Review of the Literature. *Ageing Res. Rev.* **2011,** *10*, 430–439.

Martínez, A.; Villarroel, V.; Seoane, J.; Pozo, F. D. Rural Telemedicine for Primary Healthcare in Developing Countries. *IEEE Technol. Soc. Mag.* **2004,** *23* (2), 13–22.

Mihova, P.; J. Vinarova, A. Petkov, I. Penjurov Milestone before/after Analysis of Telemedicine Implementation. *Ukrain. J. Telemed. Med. Telemat.: Sci. Pract. J.* **2009,** *7* (1), 65–67.

NHS. *The NHS Long Term Plan*, 2019. https://www.longtermplan.nhs.uk/wp-content/uploads/2019/01/nhs-long-term-plan-june-2019.pdf (accessed January 07, 2021).

Noel, H. C.; Vogel, D. C.; Erdos, J. J.; Cornwall, D.; Levin, F. Home Telehealth Reduces Healthcare Costs. *Telemed. J. E. Health* **2004,** *10*, 170–183.

Omboni, S., McManus, R.J.; Bosworth, H. B.; Chappell, L. C.; Green, B. B.; Kario, K.; Logan, A. G.; Magid, D. J.; Mckinstry, B., Margolis, K. L.; Parati, G. Evidence and Recommendations on the Use of Telemedicine for the Management of Arterial Hypertension: An International Expert Position Paper. *Hypertension* **2020,** *76* (5), 1368–1383.

PACITA Project. EU Stakeholder Involvement on Ageing Society. Telecare Technology for an Ageing Society in Europe. *Current State and Future Developments*, 2014. Available from: http://wp6.pacitaproject.eu/wp-content/uploads/2014/02/Telecare-description-web.pdf (accessed December 29, 2014).

Parè, G.; Jaana, M.; Sicotte, C. Systematic Review of Home Telemonitoring for Chronic Diseases: The Evidence Base. *J. Am. Med. Inform. Assoc.* **2007,** *14*, 269–277.

Peek, N.; Sujan, M.; Scott, P. Digital Health and Care in Pandemic Times: Impact of COVID-19. *BMJ Health Care Inform.* **2020,** *27* (1), e100166.

Pooni, R.; Pageler, N. M.; Sandborg, C.; Lee, T. Pediatric Subspecialty Telemedicine Use from the Patient and Provider Perspective. *Pediatr. Res.* **2022,** *91*, 241–246.

Rechel, B.; Grundy, E.; Robine, J. M.; et al. Ageing in the European Union. *Lancet* **2013,** *381*, 1312–1322.

Rockwell, K. L.; Gilroy, A. S. Incorporating Telemedicine as Part of COVID-19 Outbreak Response Systems. *Am. J. Manage. Care* **2020,** *26* (4), 147–148.

Rosenthal, T. C.; Williams, M. E.; Naughton, B. J., Eds. *Office Care Geriatrics*, 6th ed. Lippincott Williams & Wilkins: Philadelphia, PA, USA, 2006.

Satou, G. M.; Rheuban, K.; Alverson, D.; Lewin, M.; Mahnke, C.; Marcin, J.; et al. Telemedicine in Pediatric Cardiology: A Scientific Statement from the American Heart Association. *Circulation* **2017,** *135*, e648–e678. doi:10.1161/CIR.0000000000000478.

Sharma, R.; Fleischut, P.; Barchi, D. Telemedicine and Its Transformation of Emergency Care: A Case Study of One of the Largest US Integrated Healthcare Delivery Systems. *Int. J. Emerg. Med.* **2017,** *10*, 21. doi:10.1186/s12245-017-0146-7.

Shi, S.; Qin, M.; Shen, B.; Cai, Y.; Liu, T.; Yang, F.; et al. Association of Cardiac Injury with Mortality in Hospitalized Patients with COVID-19 in Wuhan, China. *JAMA Cardiol.* **2020,** *5*, 802–810. doi:10.1001/jamacardio.2020.0950.

Solli, H.; Hvalvik, S.; Bjørk, I. T.; Hellesø, R. Characteristics of the Relationship That Develops from Nurse-Caregiver Communication During Telecare. *J Clin Nurs.* **2015,** *24*, 1995–2004.

Styles, V. Service Users' Acceptability of Videoconferencing as a Form of Service Delivery. *J. Telemed. Telecare* **2008,** *14* (8), 415–420.

Takahashi, P. Y.; Hanson, G. J.; Pecina, J. L.; et al. A Randomized Controlled Trial of Telemonitoring in Older Adults with Multiple Chronic Conditions: The Tele-ERA Study. *BMC Health Serv. Res.* **2010,** *10*, 255.

Tchero, H.; Tabue Teguo, M.; Lannuzel, A.; Rusch, E. Telerehabilitation for Stroke Survivors: Systematic Review and Meta-analysis. *J. Med. Internet Res.* **2018,** *20* (10), e10867.

Tenforde, A. S.; Borgstrom, H.; Polich, G.; Steere, H.; Davis, I. S.; Cotton, K.; et al. Outpatient Physical, Occupational, and Speech Therapy Synchronous Telemedicine: A Survey Study of Patient Satisfaction with Virtual Visits during the COVID-19 Pandemic. *Am. J. Phys. Med. Rehabil.* **2020,** *99* (11), 977–981.

Tomita, M. R.; Russ, L. S.; Sridhar, R.; Naughton, B. J. Smart Home with Healthcare Technologies for Community-Dwelling Older Adults. In *Smart Home Systems*; Al-Qutayr, M. A., Ed.; InTech, 2010. Available from: http://www.intechopen.com/books/smart-home-systems/smart-home

United States Demographics Profile. *United States Demographics Profile 2014*, 2014. Available from: http://www.indexmundi.com/united_states/demographics_profile.html (accessed December 15, 2014).

US Census Bureau. *65+ in the United States*, 2010; US Government Printing Office: Washington, DC, USA; 2014; pp 23–212. Available from: http://www.census.gov/content/dam/Census/library/publications/2014/demo/p23-212.pdf (accessed December 20, 2014).

Wade, D. T. The Future of Rehabilitation in the United Kingdom National Health Service: Using the COVID-19 Crisis to Promote Change, Increasing Efficiency and Effectiveness. *Clin. Rehabil.* **2021,** *35* (4), 471–480.

Wernhart, A.; Gahbauer, S.; Haluza, D. eHealth and Telemedicine: Practices and Beliefs among Healthcare Professionals and Medical Students at a Medical University. *PLoS One* **2019,** 14 (2), e0213067.

Wilson, L. S.; Maeder, A. J. Recent Directions in Telemedicine: Review of Trends in Research and Practice. *Healthcare Inform. Res.* **2015,** *21* (4), 213.

World Health Organization. *Ageing and Life Course. Ten Facts on Ageing and the Life Course*; World Health Organization, 2014. Available from: http://www.who.int/features/factfiles/ageing/ageing_facts/en/index.html (accessed December 15, 2014).

World Health Organization (WHO). *Artificial Intelligence for Good Global Summit.* Director Generals, 2018. Available online at: https://www.who.int/dg/speeches/2018/artificial-intelligence-summit/en/ (accessed April 12, 2020).

World Health Organization. The Impact of the COVID-19 *Pandemic on Noncommunicable Disease Resources and Services: Results of a Rapid Assessment 2020*; World Health

Organization, 2020. https://www.who.int/publications/i/item/ncds-covid-rapid-assessment (accessed November 17, 2021).

Yalon-Chamovitz, S.; Weiss, P. L. Virtual Reality as a Leisure Activity for Young Adults with Physical and Intellectual Disabilities. *Res. Dev. Disabil.* **2008,** *29* (3), 273–287.

Yellowlees, P. M.; Chorba, K.; Burke Parish, M.; Wynn-Jones, H.; Nafiz, N. Telemedicine Can Make Healthcare Greener. *Telemed. e-Health* **2010,** *16* (2), 229–232.

CHAPTER 18

ALTERNATIVE TREATMENT AND MEDICINES FOR TRAVELERS

ABSTRACT

Travelers from across the world are becoming increasingly cognizant of the alternative treatments and complementary medicine available to address their conditions and ailments, prompting increased wellness tourism to countries with ancient or traditional alternative treatments such as China, Korea, and India. From Ayurveda to yoga to naturopathy, this chapter explores how the international perception of wellness is shifting alongside the global rise in popularity of these alternative medical treatments.

Everywhere in the world, there is fierce competition in the healthcare sector. In the past, wealthy patients from poor nations frequently traveled to developed nations for top-notch medical care (Herrick, 2007). Today, a lot of underprivileged people from rich nations come to what was once considered "third world" regions for medical treatment. These individuals are seeking cheap, high-quality medical care. Many countries around the world, especially India, are becoming popular treatment destinations as a result.

Medical travelers from all over the world are increasingly drawn to complementary and alternative medicine (CAM). CAM is a collection of various medical and healthcare practices, systems, and goods that are not currently seen as belonging to traditional medicine. As CAM therapies and techniques that have been shown to be safe and effective are being adopted as "mainstream" healthcare practices, the list of procedures that are regarded as CAM is constantly changing. CAM procedures today can be divided into five primary categories: CAM, mind-body interventions, physiologically based therapies, manipulative and body-based techniques, and energy therapies.

Medical, Dental, and Wellness Tourism: A Postpandemic Perspective. Mary Schreiber Swenson, PhD, and Amit Bansal (Authors).

The ancient medical systems of China, Korea, Japan, India, and Vietnam each have their own distinctive variations. Under the general heading of "complementary and alternative medicine," alternative medicine is frequently lumped along with complementary medicine. In contrast to alternative medicine, which is used in place of conventional medicine, complementary medicine refers to treatments that support standard western (or allopathic) medicine. Alternative medicine describes therapeutic modalities used in place of conventional medicine to treat ailments. Integrative medicine refers to the blending of alternative therapies with traditional medical care. Holistic care, which emphasizes treating a person as a whole, is a fundamental component of CAM.

Ayurveda is an example of CAM therapeutic systems. It has its roots in India and dates back more than 5000 years. It emphasizes a specific treatment depending on the conditions of each patient. In addition to yoga, meditation, massage, diet, and herbal remedies, homeopathy employs very small dosages of a drug that creates symptoms in order to trigger the body's natural healing process. Naturopathy focuses on noninvasive procedures to support your body's natural healing processes. Chinese, Asian, Pacific Islander, American Indian, and Tibetan techniques are examples of traditional complementary and alternative therapies.

While there are some similarities among the various medical systems, they all have different philosophical perspectives on disease prevention and treatment. These approaches are predicated on the idea that the body is capable of healing itself. Utilizing a variety of methods that incorporate the mind, body, and spirit is frequently necessary for healing. Depending on the symptoms that are currently present, treatment is frequently customized.

Ayurveda is acknowledged as traditional Indian medicine, and there has been a rise in interest in traditional medicine globally in recent years (Dawn and Pal, 2011; Kunwar, 2019; Patwardhan et al., 2005). The conventional medical system has expanded globally due to the contemporary epidemiological transition (Gupta et al., 2015), and more people are crossing international borders to access cutting-edge medical care (Dawn and Pal, 2011). The traditional medicinal treatment was invented and is well-known in India, as are wellness tourism and alternative medical care. Yoga and Ayurveda are currently used interchangeably with Indian medicine. The Indian market is all set to grow in alternative treatments as well as wellness tourism in the coming years as well.

Mishra and Shailesh (2012) researched alternative medical care and found that when patients travel abroad, their priority is to receive top-notch care. The Indian healthcare system is renowned for offering high-quality medical care at a reasonable price, according to Gupta et al. (2015).

In India, the practice of alternative medicine, particularly AYUSH (Ayurveda, Yoga and Naturopathy, Unani, Siddha, and Homeopathy), is growing in popularity. The Indian government established the Ministry of AYUSH in 2014, in part to encourage medical travel to the nation, reflecting the rise in interest. The government has maintained to raise the budget for the AYUSH Ministry each year in order to implement appropriate strategies; in 2018, the total budget increased by more than 13% compared to the prior year [PTI, 2018]. To create a skilled workforce for the medical industry, the ministry has committed a sizable percentage of its funding to research and education (Ministry of AYUSH, 2019). Additionally, a central committee has been established to develop the essential framework and implement a strong regulatory framework.

The demand for alternative medical treatments is rising quickly as alternative medical systems are becoming more widely known. In 2014, the AYUSH market was about $3 billion; today, it is worth more than $18 billion. With projects involving the repackaging of medications and formulations for Ayurveda and homeopathy as well as the provision of preventive and holistic treatments, numerous startups are attempting to increase this industry. These alternative medical treatments are advised for patients who have had negative effects from allopathic medications or have failed to respond to them.

18.1 ALTERNATIVE MEDICAL TREATMENTS OF AYUSH

The AYUSH Ministry of India was founded in November 2014 to guarantee the greatest possible growth and diffusion of AYUSH systems. The ministry was established with the goal of enhancing instruction and fostering research in AYUSH (Ministry of AYUSH, 2019). The ministry has been attracting medical travelers from corners of the world since its establishment. The abbreviation AYUSH stands for Ayurveda, Yoga and Naturopathy, Unani, Siddha, and Homoeopathy, as previously mentioned. Below is a description of each:

Ayurveda: Ayurveda offers a comprehensive system of healthcare and preventive medicine. It has a long history of success. Ayurveda is based on organic herbs that have benefits for each individual. Ayurveda is a science that is built on understanding how people live. The five elements—ether, air, fire, water, and earth—make up the universe, and it is the most crucial fundamental idea in Ayurveda. They are displayed in three biological Dashas named Vata, Pitta, and Kapha (Dawn and Pal, 2011). The individuals going through this treatment fully comprehend all facets of bodily function and create the harmonious balance required for a long and healthy life.

Yoga and Naturopathy: It is right to say that Yoga has gained the attention of the world. It is a practice that helps people develop their innate strength in a healthy way. It offers a way to obtain total self-fulfillment. Yoga is the control of mental shifts (AYUSH, 2018). It is a tool for enlightenment and self-evolution via physical and spiritual health. A variety of yoga poses are necessary for organs to maintain their optimal function. Yoga balances the internal organs, cleanses and detoxifies the body, and strengthens the immune system. According to naturopathy, the human body possesses special capacities for self-healing and self-building. It examines and addresses all of the factors responsible for these conditions, such as abnormal ways of living, thinking, sleeping, and working, as well as environmental factors that can impair the body's ability to function normally, rather than focusing on the precise cause of a disease and its specific treatment.

Unani: According to the Unani Medical System, the four fundamental elements of earth, air, water, and fire—along with their associated temperaments of cold, hot, wet, and dry—make up the human body. When these components work together and interact, a new compound with a new character—namely, hot-wet, hot-dry, cold-wet, and cold-dry—is created. The Unani medical system is based on physical examinations and observation. The disease of a person should be seen as their own substance and creativity. The Unani system advocates for healthy living, disease prevention, and treatment. Numerous procedures are part of Unani medicine. Venesection, cupping, Turkish baths, massage, cauterization, purging, exercise, leeching, and other practices are all part of Unani therapy. Unani diet therapy provides particular diets or controls the type and amount of food consumed. Unani pharmacology makes use of primarily herbal, natural remedies. Unani physiotherapy employs specific workout methods to help the body's homeostasis remain in balance (World Health Organization, 2010).

Siddha: According to the Siddha system, an illness is a condition in which the five elements' regular equilibrium is lost, resulting in various degrees of discomfort and anguish. The Siddha technique of therapy is based on the doctor's clinical judgment following observation of the patient and examination of the pulse, diagnosis, and medical background. Siddha claims that specific diets can boost physical and mental function or even lower the chance of sickness. "Food Is Medicine and Medicine Is Food," is the central tenet of this traditional Indian medical practice. Along with prescribed medications for every ailment, certain meal dishes and grains are also given (Sivaraman and Rajalakshmi, 2005).

Homeopathy: Today, homeopathy is a fast-expanding medical system used throughout the globe. It is well known for both the tenderness of its treatment and the safety of its medicines. In India, almost 10% of the population relies on homeopathy for their medical requirements, according to one survey. It is regarded as India's second-most well-liked medical system (Ministry of AYUSH, 2019). The patient with their condition and personal traits, the medications with their composition and production procedures, and the doctors with the various treatment modalities and health philosophies are the three primary agents in homeopathic treatments (Poitevin, 1999).

All of the above-mentioned alternative medical systems draw patients from throughout the globe and promote medical tourism in India.

18.2 UNDERSTANDING THE WELLNESS TOURISM

Activities and treatments designed to balance and improve all the major facets of human life are included in wellness tourism. Wellness encompasses all facets of a person's way of life, including the physical, mental, emotional, occupational, intellectual, and spiritual facets. The primary goal of wellness travel is to engage in preventive and active lifestyle enhancement activities, such as healthy food, exercise, rest, luxurious living, and therapeutic treatments (World Tourism Organization, 2019). Wellness tourism frequently includes things like spa services, stress reduction, anti-aging facilities, meditation, acupressure, and acupuncture.

The Global Wellness Institute (2017) reported that weight loss, nutrition, healthy living, a wellness lifestyle, mineral springs, and the hot springs are the most important aspects of wellness tourism. The United States, Germany, France, China, and Japan are among the top nations for wellness travel. The number of wellness tourism visits to India rose by 18% between 2013 and 2015 (Global Wellness Institute, 2016), while the industry's earnings have increased by 7.8% over the previous five years (Global Wellness Institute, 2018). India ranked seventh among the top 20 wellness tourist markets in 2017, according to estimates from Global Wellness Tourism, bringing in a total of USD 16.3 billion. By 2022, wellness tourism is expected to grow at an average annual rate of 7.5%.

18.3 CHINESE TRADITIONAL MEDICINE

One of the oldest alternative medicines in the world is traditional Chinese medicine (TCM). Its roots go back centuries. In traditional Chinese books, there is evidence of treating many diseases like rheumatoid arthritis and cancer with the help of traditional medicine. The foundation of TCM is the idea that all natural phenomena can be characterized as being yin yang (yn yáng), a duality of complementary forces, and that these forces correspond to the five phases, wood, fire, earth, metal, and water that cause changes and movements. "Well-being" is thought to result from both an exterior harmony with one's social and natural environment and an internal harmony of the body. This balance is based on the interdependence of yin and yang. For example, illnesses exhibiting strength, forceful movements, coldness, and under activity are comparatively more yin, whereas illnesses exhibiting overactivity, forceful movements, and strength are comparatively more yang.

Because yin has qualities that are more yang and vice versa, every yin yang classification can be divided indefinitely. For instance, yin disorders typified by coolness could also display yang traits like convulsions that come on suddenly. Yin and yang are dynamic forces that adapt to their environment rather than remaining constant. A person's health and welfare are determined by this constantly changing equilibrium.

Treatments used in TCM work to restore and maintain balance. For instance, nourishing yin is used to treat yin depletion. Rhizomes from common anemarrhena plants (*Anemarrhena asphodeloides*) and Reeve's turtle shells (*Mauremys reevesii*) both nourish yin, although their applications have different circumstances (Hijikata et al., 2006; Xie and Du, 2011). This is due to the fact that each medicinal ingredient's distinctive features restrict their usage to certain patterns of disharmony, which are imbalances particular to the patient as a whole person (Jiang et al., 2012).

Even when one is healthy, TCM places a strong emphasis on maintaining balance. Medical intervention is only used to treat discord when less drastic approaches fail (Koh, 2010; Wilms, 2010). Yangsheng (/yang sheng, nourishing life) is a practice that makes minor adjustments to maintain equilibrium (Liu et al., 2013). Because some therapeutic components are ingested as food or tonics to improve general health, their use is not always dictated by an immediate medical need (Chau and Wu, 2006; Koo, 1984; Smith, 2018). For instance, tiger bone (/h/g; *Panthera tigris*) is processed into wine and consumed to strengthen bones in addition to being used to cure rheumatic pain (Bensky et al., 2004; Gratwicke et al., 2008; Wong, 2016).

Through the use of lunzhi, circular finger motions (Jiang et al., 2012), practitioners of TCM choose therapies for their patients. Pharmaceutical therapy, acupuncture, moxibustion, therapeutic massage, dietary therapy, and physical activity are just a few of the many treatments used in TCM (Critchley et al., 2000; Griffiths et al., 2010). A single TCM treatment can treat a variety of biomedically defined disorders by viewing sickness as the manifestation of structured disharmony (Kaptchuk, 2000). Biomedicine treats liver fibrosis, biliary cirrhosis, and liver cancer quite differently, although TCM patent medicines have been used to treat all three diseases by reducing liver fire and heat (Li et al., 2016).

While TCM's standing has improved globally, its use for treating refractory illnesses has grown more widespread in China and other Asian nations. For arthritis, musculoskeletal disorders, weariness, and health maintenance, about 75% of Chinese or immigrant Chinese patients prefer TCM or other complementary and alternative therapies (Manheimer et al., 2009), and this percentage is on a rise in the west. Over the past few decades, many pieces of research have been carried out in China to hasten the transition of TCM from an experience-based medicine to evidence-based medicine; these studies have significantly improved the acceptability of TCM therapies and the scientific reputation of TCM globally attracting more and more patients from around the world. With the ever-growing curiosity, TCM is suspected to stay on a steep rise in the coming years.

18.4 FUTURE POSSIBILITIES OF ALTERNATIVE AND COMPLEMENTARY MEDICAL TREATMENT

In recent years, there has been a surge in the number of patients looking for alternative medical treatments owing to enhanced computerization and the spread of the mass media (Suzuki, 2004). A growing trend in medicine is alternative medicine. Traditional medicine and folk medicines from different nations and regions of the world are included in the wide variety of alternative medicine, along with innovative treatments that are not covered by health insurance. Surprisingly, only a few individuals believe that contemporary western medicine is beneficial; in fact, the World Health Organization (WHO) labels 65–80% of its global healthcare activities as "traditional medicine" (Suzuki, 2004). The growing acceptance of alternative medical practices has many justifications. These therapies and treatments make the patient feel better, encourage relaxation, and lower stress levels, enabling those with serious illnesses like cancer to manage their condition more effectively.

There is significant evidence that some complementary therapies can manage or even cure some cancer symptoms while lowering the severe adverse effects of powerful drugs. For instance, acupuncture can help with the discomfort and illness brought on by some chemotherapy medications. After head and neck cancer therapy, it can also aid in easing mouth pain.

Additionally, for some patients, a substantial benefit is the comfort and happiness that result from the touch, conversation, and time that a complementary therapist typically provides. Having the best therapist at your side can be beneficial for your recovery from cancer therapy. For instance, a patient may feel empathetic after receiving expert and considerate aromatherapy, which may aid in the patient's recuperation. Some patients turn to complementary therapies in order to feel upbeat and cheerful, emotions that a doctor would find challenging to instill but that alternative medical therapies might be able to encourage.

According to several studies, a significant portion of the senior population—between 35 and 60%—uses CAM, which is greater than the proportion for other age groups (Astin, 1998). Approximately 841 million people in the globe were 60 years of age or older, making up more than 60% of the overall population, according to a 2013 European Environment Agency assessment (European Environment Agency, 2013). Between 2015 and 2050, the proportion of the world's population over 60 years will nearly double from 12% to 22% (WHO). Nearly 8 out of 10 of these old people are anticipated to reside in less developed nations by 2050 (Crimmins, 2004).

The most often used therapy approach for senior individuals right now is Ayurveda, the most popular alternative system practiced by Indians; homeopathy is another well-liked Indian treatment system (Sharma et al., 2017). The National Health Interview Survey from 2007 found that yoga and other mind-body therapies were the most popular CAM treatments for kids with behavioral, emotional, or mental health issues.

These statistics highlight the potential for alternative tourism globally, and by 2022, its market is projected to be more than treble. In comparison to the marketplaces in most Western countries, the rebirth of traditional and alternative medicine, such as AYUSH (IANS, 2017), TCMs, and Turkish spring baths has boosted the medical tourism sector.

KEYWORDS

- **alternative treatment**
- **naturopathy**
- **wellness tourism**
- **travel for treatment**
- **medical tourism**
- **Ayurveda**
- **complementary medicine**

REFERENCES

Astin, J. Why Patients Use Alternative Medicine: Results of a National Study. *JAMA* **1998,** *279* (19), 1548–1553. doi: 10.1001/jama.279.19.1548.

Bensky, D.; Clavey, S.; Stöger, E.; Gamble, A. *Chinese Herbal Medicine: Materia Medica.* Eastland Press Inc.: Seattle, WA, 2004.

Chau, C. F.; Wu, S. H. The Development of Regulations of Chinese Herbal Medicines for Both Medicinal and Food Uses. *Trends Food Sci. Technol.* **2006,** *17*, 313–323. doi: 10.1016/j.tifs.2005.12.005.

Crimmins, E. Trends in the Health of the Elderly. *Annu. Rev. Public Health* **2004,** *25* (1), 79–98. doi: 10.1146/annurev.publhealth.25.102802.124401.

Critchley, J. A. J. H.; Zhang, Y.; Suthisisang, C. C.; Chan, T. Y. K.; Tomlinson, B. Alternative Therapies and Medical Science: Designing Clinical Trials of Alternative/Complementary Medicines—Is Evidence-Based Traditional Chinese Medicine Attainable? *J. Clin. Pharmacol.* **2000,** *40*, 462–467.

Dawn, S.; Pal, S. Medical Tourism in India: Issues, Opportunities and Designing Strategies for Growth and Development. *Int. J. Multidiscipl. Res.* **2011,** *1* (3), 185–202.

European Environment Agency. *Increasing Divergence in Global Population Trends: Analysis for Update and Improved Assessment of This Megatrend*; European Environment Agency, 2013; pp 9–11.

Global Wellness Institute. 2016. Retrieved August 21, 2019, from https://globalwellnessinstitute.org/press-room/press-releases/international-wellness-tourism-growingmuch-faster-than-domestic/

Global Wellness Institute. *Global Wellness Economy Monitor*; Global Wellness Institute, 2017; pp 3–20.

Global Wellness Institute. *Global Wellness Economy Monitor*; Global Wellness Institute, 2018; pp 3–9.

Gratwicke, B.; Mills, J.; Dutton, A.; Gabriel, G.; Long, B.; Seidensticker, J.; Wright, B.; You, W.; Zhang, L. Attitudes toward Consumption and Conservation of Tigers in China. *PLoS ONE* **2008,** *3*, e2544.

Griffiths, S. M.; Chung, V. C. H.; Tang, J. L. Integrating Traditional Chinese Medicine: Experiences from China. *Australas. Med. J.* **2010,** *3*, 385–396.

Gupta, M.; Rajachar, V.; Prabha, C. Medical Tourism: A New Growth Factor for Indian Healthcare Industry. *Int. J. Res. Med. Sci.* **2015,** 2161–2163. doi: 10.18203/2320-6012.ijrms20150597.

Herrick, D. Medical Tourism: Global Competition in Health Care. *National Center for Policy Analysis (NCPA) Policy Report No. 304*, 2007.

Hijikata, Y.; Yasuhara, A.; Yoshida, Y.; Sento, S. Traditional Chinese Medicine Treatment of Epilepsy. *J. Altern. Complement. Med.* **2006,** *12*, 673–677.

IANS. *AYUSH Boosting Medical Tourism in India: Experts—ET HealthWorld*, 2017. Retrieved August 20, 2019, from https://health.economictimes.indiatimes.com/news/industry/ayush-boosting-medicaltourism-in-india-experts/59725359

Jiang, M.; Lu, C.; Zhang, C.; Yang, J.; Tan, Y.; Lu, A.; Chan, K. Syndrome Differentiation in Modern Research of Traditional Chinese Medicine. *J. Ethnopharmacol.* **2012,** *140*, 634–642.

Kaptchuk, T. J. *Chinese Medicine: The Web That Has No Weaver*, Rev ed.; Rider: London, 2000.

Koh, A. Wen Bing (Warm Diseases) and the 2009 H1N1 Influenza. *Austral. J. Acupunct. Chin. Med.* **2010,** *5*, 23–29.

Koo, L. C. The Use of Food to Treat and Prevent Disease in Chinese Culture. *Soc. Sci. Med.* **1984,** *18*, 757–766. doi: 10.1016/0277-9536(84)90102-3.

Kunwar, R. Medical Tourism and Hospitality in Hospital. *Gaze: J. Tour. Hosp.* **2019,** *10* (1), 67–123. doi: 10.3126/gaze.v10i1.22778.

Li, S.; Tan, H. Y.; Wang, N.; Hong, M.; Li, L.; Cheung, F.; Feng, Y. Substitutes for Bear Bile for the Treatment of Liver Diseases: Research Progress and Future Perspective. *Evid.-Based Complement. Altern. Med.* **2016,** *2016*, 4305074. doi: 10.1155/2016/4305074.

Liu, J.; Pei, M.; Zheng, C.; Li, Y.; Wang, Y.; Lu, A.; Yang, L. A Systems-Pharmacology Analysis of Herbal Medicines Used in Health Improvement Treatment: Predicting Potential New Drugs and Targets. *Evid.-Based Complement. Altern. Med.* **2013,** *2013*, 17.

Manheimer, E.; Wieland, S.; Kimbrough, E.; Cheng, K.; Berman, B. M. Evidence from the Cochrane Collaboration for Traditional Chinese Medicine Therapies. *J. Altern. Complement. Med.* **2009,** *15* (9), 1001–1014.

Ministry of AYUSH. *Annual Report*; Government of India Ministry of Ayurveda, Yoga & Naturopathy, Unani, Siddha, Sowa-Rigpa, and Homeopathy (AYUSH), 2019. Retrieved August 14, 2019, from http://ayush.gov.in/about-us/about-theministry

Mishra, D.; Shailesh, K. Making Indian Healthcare Market a Global Medical Tourism Destination. *IOSR J. Bus. Manage.* **2012,** *2* (4), 23–28. doi: 10.9790/487x-0242328.

Poitevin, B. Integrating Homoeopathy in Health Systems. *Bull. World Health Organ.* **1999,** *77* (2), 160–166.

Press Trust of India (PTI). *AYUSH Ministry Allocation in Budget Increased by 13 Per Cent*, 2018. Retrieved September 3, 2019, from https://timesofindia.indiatimes.com/business/india-business/ayush-ministry-allocation-in-budgetincreased-by-13-per-cent/articleshow/62743094.cms

Sharma, E.; Dubey, A.; Malhotra, S.; Manocha, S.; Handu, S. Use of Complementary and Alternative Medicines in Indian Elderly Patients. *Natl. J. Physiol., Pharm. Pharmacol.* **2017,** *7* (9), 1. doi: 10.5455/njppp.2017.7.0411401052017.

Sivaraman, G.; Rajalakshmi, K. Concept of Siddha Nutraceuticals. *Acta Hortic.* **2005,** *6*, 153–159. doi: 10.17660/actahortic.2005.680.23.

Smith, A. A. *Capturing Quicksilver—The Position, Power, and Plasticity of Chinese Medicine in Singapore*; Berghahn Books: New York, NY, 2018.

Suzuki, N. Complementary and Alternative Medicine: A Japanese Perspective. *Evid.-Based Complement. Altern. Med.* **2004,** *1* (2), 113–118. doi: 10.1093/ecam/neh029.

Wilms, S. Nurturing Life in Classical Chinese Medicine: Sun Simiao on Healing without Drugs, Transforming Bodies and Cultivating Life. *J. Chin. Med.* **2010,** *93*, 5–13.

Wong, R. W. Y. The Organization of the Illegal Tiger Parts Trade in China. *Br. J. Criminol.* **2016,** *56*, 995–1013.

World Health Organization. Benchmarks for Training in Traditional/Complementary and Alternative Medicine. *WHO Benchmarks for Training in Unani Medicine*; WHO, 2010; pp 1–17.

World Tourism Organization. *Implementation of the General Programme of Work for 2016–2017—CE/106/3*; (UNWTOA) Specialized Agency of the United Nations, 2019.

Xie, W.; Du, L. Diabetes Is an Inflammatory Disease: Evidence from Traditional Chinese Medicines. *Diab., Obes. Metab.* **2011,** *13*, 289–301.

INDEX

D

E

H

I

J

K

L

M